GRAPHING CALCULATOR MANUAL

BEVERLY FUSFIELD

COLLEGE ALGEBRA: GRAPHS AND MODELS

FIFTH EDITION

Marvin L. Bittinger

Indiana University Purdue University Indianapolis

Judith A. Beecher

Indiana University Purdue University Indianapolis

David J. Ellenbogen

Community College of Vermont

Judith A. Penna

Indiana University Purdue University Indianapolis

PEARSON

Boston   Columbus   Indianapolis   New York   San Francisco   Upper Saddle River
Amsterdam   Cape Town   Dubai   London   Madrid   Milan   Munich   Paris   Montreal   Toronto
Delhi   Mexico City   Sao Paulo   Sydney   Hong Kong   Seoul   Singapore   Taipei   Tokyo

Copyright © 2013, 2009, 2005 Pearson Education, Inc.
Publishing as Pearson, 75 Arlington Street, Boston, MA 02116.

ISBN-13: 978-0-321-79099-6
ISBN-10: 0-321-79099-5

2 3 4 5 6 EBM 15 14 13 12

www.pearsonhighered.com

Table of Contents

Preface i

Chapter R Basic Concepts of Algebra 1

Chapter 1 Graphs, Functions, and Models 9

Chapter 2 More on Functions...23

Chapter 3 Quadratic Functions and Equations; Inequalities 27

Chapter 4 Polynomial Functions and Rational Functions.............................31

Chapter 5 Exponential Functions and Logarithmic Functions......................35

Chapter 6 Systems of Equations and Matrices..45

Chapter 7 Conic Sections 63

Chapter 8 Sequences, Series, and Combinatorics 73

Chapter R Basic Concepts of Algebra

GETTING STARTED

Before turning on the calculator, note that there are options above the keys as well as on them. To access the option written on a key, simply press the key. The options written to the left above the keys are accessed by first pressing the [2nd] key in the left column of the keypad and then pressing the key corresponding to the desired option. These options are labeled in blue on the TI-84 models. The options written in green to the right above the keys are accessed by first pressing the green [ALPHA] key in the left column of the keypad.

Press [ON] to turn on the calculator. ([ON] is the key at the bottom left-hand corner of the keypad.) You should see a blinking rectangle, or cursor, on the screen. If you do not see the cursor, try adjusting the display contrast. To do this, first press [2nd]. Then press and hold [▲] to increase the contrast or [▼] to decrease the contrast. If the contrast needs to be adjusted further after the first adjustment, press [2nd] again and then hold [▲] or [▼] to increase or decrease the contrast, respectively.

To turn the calculator off, press [2nd] [OFF]. (OFF is the second operation associated with the [ON] key.) The calculator will turn itself off automatically after about five minutes without any activity.

Press [MODE] to display the MODE settings. Initially you should select the settings on the left side of the display as shown below. The Mode screen on a TI-84 Plus with MathPrint is shown below.

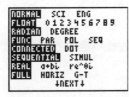

To change a setting on the Mode screen use [▼] or [▲] to move the cursor to the line of that setting. Then use [▶] or [◀] to move the blinking cursor to the desired setting and press [ENTER]. Press [CLEAR] or [2nd] [QUIT] to leave the MODE screen. (QUIT is the second operation associated with the [MODE] key.) Pressing [CLEAR] or [2nd] [QUIT] will take you to the home screen where computations are performed.

It will be helpful to read the Getting Started section of the Texas Instruments Guidebook that was packaged with your graphing calculator and the preface of this manual before proceeding.

ABSOLUTE VALUE

Section R.1, Example 3 Find the distance between −2 and 3.

The distance between −2 and 3 is $|-2 - 3|$, or $|3 - (-2)|$. Absolute value notation is denoted "abs" on the calculator. It is item 1 on the MATH NUM menu.

```
MATH NUM CPX PRB
1:abs(
2:round(
3:iPart(
4:fPart(
5:int(
6:min(
7↓max(
```

Note that the $\boxed{(-)}$ key in the bottom row of the keypad must be used to enter a negative number. The $\boxed{-}$ key in the right column of the keypad is used for the subtraction operation. Thus, to enter $|-2 - 3|$ in Classic mode press $\boxed{\text{MATH}}$ $\boxed{\blacktriangleright}$ $\boxed{\text{ENTER}}$ $\boxed{(-)}$ $\boxed{2}$ $\boxed{-}$ $\boxed{3}$ $\boxed{)}$ $\boxed{\text{ENTER}}$. To enter $|3 - (-2)|$ in Classic mode press $\boxed{\text{MATH}}$ $\boxed{\blacktriangleright}$ $\boxed{\text{ENTER}}$ $\boxed{3}$ $\boxed{-}$ $\boxed{(}$ $\boxed{(-)}$ $\boxed{2}$ $\boxed{)}$ $\boxed{)}$ $\boxed{\text{ENTER}}$. Note that the calculator supplies the left parenthesis in the absolute value notation when in Classic mode. We close the expression with a right parenthesis although it is not necessary to do so. In MathPrint mode, the calculator displays absolute value bars. Use the same series of keystrokes; however, press $\boxed{\blacktriangleright}$ instead of the last $\boxed{)}$ before pressing $\boxed{\text{ENTER}}$. In either mode, it is not necessary to include the inner set of parentheses around −2 in the second computation, but since they allow the expression to be read more easily, we include them here.

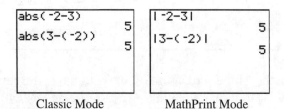

Classic Mode   MathPrint Mode

Instead of pressing $\boxed{\text{MATH}}$ $\boxed{\blacktriangleright}$ $\boxed{\text{ENTER}}$ to access "abs(" and copy it to the home screen, we could have pressed $\boxed{\text{MATH}}$ $\boxed{\blacktriangleright}$ $\boxed{1}$ since "abs(" is item 1 on the MATH NUM menu. Note that, in general, an item can be selected from a menu in two ways. Either press the number or letter shown at the left of the item or use the $\boxed{\blacktriangledown}$ or $\boxed{\blacktriangle}$ key to highlight the item and then press $\boxed{\text{ENTER}}$.

In MathPrint mode, absolute value can be accessed through a shortcut menu by pressing [ALPHA] [F2] [1] or [ALPHA] [F2] [ENTER]. (See Preface page ii for more information on shortcut menus.)

Absolute value notation can also be found as the first item in the CATALOG and copied to the home screen. To do this press [2nd] [CATALOG] [ENTER]. (CATALOG is the second operation associated with the [0] numeric key.)

EDITING ENTRIES

After you have performed a computation, you can recall and edit an entry if necessary. Suppose, for instance, in entering one of the expressions in Example 3 on page 2, you pressed 6 instead of 3. To correct this, first press [2nd] [ENTRY] to return to the last entry. (ENTRY is the second operation associated with the [ENTER] key.) Then use the [◄] key to move the cursor to 6 and press 3 to overwrite it. If you forgot to type the 2 in the first expression, move the cursor to the subtraction symbol; then press [2nd] [INS] [2] to insert the 2 before that symbol. (INS is the second operation associated with the [DEL] key.) You can continue to insert symbols immediately after the first insertion without pressing [2nd] [INS] again. If you typed 21 instead of 2, move the cursor to 1 and press [DEL]. This will delete the 1. If you notice that an entry needs to be edited before you press [ENTER] to perform the computation, the editing can be done directly without recalling the entry.

The keystrokes [2nd] [ENTRY] can be used repeatedly to recall entries preceding the last one. Pressing [2nd] [ENTRY] twice, for example, will recall the next to last entry. Using these keystrokes a third time recalls the third to last entry and so on. The number of entries that can be recalled depends on the amount of storage they occupy in the calculator's memory.

It is also possible to scroll through entries using the [▲] key. When the entry to be edited is highlighted, press [ENTER] to paste it on the current entry line.

SCIENTIFIC NOTATION

To enter a number in scientific notation, first type the decimal portion of the number, then press [2nd] [EE] (EE is the second operation associated with the [,] key.). Finally type the exponent, which can be at most two digits. For example, to enter 1.789×10^{-11} in scientific notation, press [1] [.] [7] [8] [9] [2nd] [EE] [(-)] [1] [1] [ENTER]. To enter 6.084×10^{23} in scientific notation, press [6] [.] [0] [8] [4] [2nd] [EE] [2] [3] [ENTER]. The decimal portion of each number appears before a small E while the exponent follows the E.

```
1.789E-11
            1.789E-11
6.084E23
            6.084E23
```

The graphing calculator can be used to perform computations in scientific notation. To ensure that the result will be expressed in scientific notation, select "SCI" on the first line of the MODE screen.

Section R.2, Example 8 *Distance to Neptune*. The distance from Earth to the sun is defined as 1 astronomical unit, or AU. It is about 93 million miles. The average distance from Earth to Neptune is 30.1 AUs (*Source*: Astronomy Department, Cornell University). How many miles is it from Earth to Neptune? Express your answer in scientific notation.

To solve this problem we find the product $93,000,000 \times 30.1$. Expressing the product in scientific notation, we have $\left(9.3 \times 10^7\right)(3.01 \times 10)$. Press ⑨ ⨀ ③ [2nd] [EE] ⑦ ⨯ ③ ⨀ ⓪ ① [2nd] [EE] ① [ENTER]. The result is 2.7993×10^9 miles.

```
9.3E7*3.01E1
            2.7993E9
```

ORDER OF OPERATIONS

Section R.2, Example 9 (b) Calculate $\dfrac{10 \div (8-6) + 9 \cdot 4}{2^5 + 3^2}$.

In order to divide the entire numerator by the entire denominator, we must enclose both the numerator and the denominator in parentheses. That is, we enter

$\left(10 \div (8-6) + 9 \cdot 4\right) \div \left(2^5 + 3^2\right)$. In Classic mode, press ⟮ ① ⓪ ÷ ⟮ ⑧ − ⑥ ⟯ + ⑨ ⨯ ④ ⟯ ÷ ⟮ ② ^ ⑤ + ③ x^2 ⟯ [ENTER]. Note that 3^2 can be entered either as ③ x^2 or as ③ ^ ②. In MathPrint mode, press ⟮ ① ⓪ ÷ ⟮ ⑧ − ⑥ ⟯ + ⑨ ⨯ ④ ⟯ ÷ ⟮ ② ^ ⑤ ▶ + ③ x^2 ⟯ [ENTER].

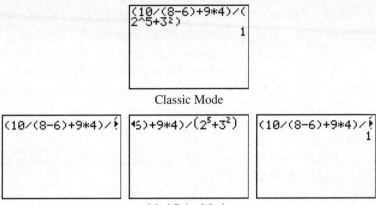

Classic Mode

MathPrint Mode

It is also possible to use the fraction shortcut to enter the expression. In MathPrint mode, press [ALPHA] [F1] [1] to paste the fraction template on the home screen. Make sure the cursor is in the numerator, then press [1] [0] [÷] [(] [8] [−] [6] [)] [+] [9] [×] [4] [▸] [2] [^] [5] [▸] [+] [3] [x^2] [▸] [ENTER].

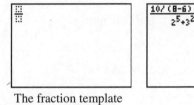 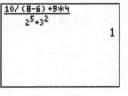

The fraction template
in MathPrint Mode

THE TVM SOLVER

Section R.2, Example 10 *Compound Interest.* If a principal P is invested at an interest rate r, compounded n times per year, in t years it will grow to an amount A given by

$$A = P\left(1 + \frac{r}{n}\right)^{nt}.$$

Suppose that $1250 is invested at 4.6% interest, compounded quarterly. How much is in the account at the end of 8 years?

To find the desired value, we could use a calculator to do the computation shown on page 14 of the text or we could use the calculator's TVM Solver. Here we will illustrate the use of the TVM Solver. This application is accessed by pressing [APPS] [ENTER] or [APPS] [1] to select the Finance application and then press [ENTER] or [1] to select the TVM Solver.

Enter $4 \cdot 8$, or 32 (the total number of compounding periods) for N. Then use the ⬇ key to position the cursor beside I% and enter 4.6 (the interest rate). Continuing in the same manner, enter PV = −1250 (the amount invested, entered as a negative number), PMT = 0, P/Y = 4, and C/Y = 4 (the number of compounding periods each year). The END/BEGIN setting at the bottom of the screen is irrelevant in this situation.

Position the cursor beside FV = and press [ALPHA] [SOLVE] to display the desired value. (SOLVE is the Alpha operation associated with the [ENTER] key.) We see that there will be $1802.26 in the account at the end of 8 years.

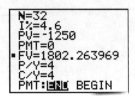

RADICAL NOTATION

We can use the square-root, cube-root, and xth-root features to simplify radical expressions.

Section R.7, Example 1 Simplify each of the following.

a) $\sqrt{36}$ b) $-\sqrt{36}$ c) $\sqrt[3]{-8}$ d) $\sqrt[5]{\dfrac{32}{243}}$ e) $\sqrt[4]{-16}$

a) To find $\sqrt{36}$, press [2nd] [√] [3] [6] [)] [ENTER]. ([√] is the second operation associated with the [x^2] key.) Note that the calculator supplies a left parenthesis with the radical symbol in Classic mode and, although it is not necessary to do so, we close the expression with a right parenthesis for completeness. Since MathPrint mode does not supply a left parenthesis, it is not necessary to press [)] in MathPrint mode.

b) To find $-\sqrt{36}$, in Classic mode press [(-)] [2nd] [√] [3] [6] [)] [ENTER]. It is not necessary to press in [)] MathPrint mode.

```
√(36)
              6
-√(36)
             -6
```
Classic Mode

```
√36
              6
-√36
             -6
```
MathPrint Mode

c) We will use the cube-root feature from the MATH menu to find $\sqrt[3]{-8}$. Press [MATH] [4] [(-)] [8] [)] [ENTER]. It is not necessary to press in [)] MathPrint mode.

d) We will use the xth-root feature from the MATH menu to find $\sqrt[5]{\dfrac{32}{243}}$. We will also use ▶Frac to express the result as a fraction. In Classic Mode, press [5] [MATH] [5] [(] [3] [2] [÷] [2] [4] [3] [)] [MATH] [1] [ENTER]. The first 5 is the index of the radical, and the second 5 is used to select item 5, the xth-root, from the MATH menu. In MathPrint mode, press [5] [MATH] [5] [3] [2] [÷] [2] [4] [3] [▶] [MATH] [1] [ENTER].

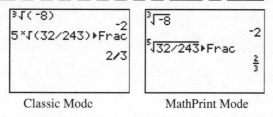

Classic Mode   MathPrint Mode

e) To enter $\sqrt[4]{-16}$ press [4] [MATH] [5] [(-)] [1] [6] [ENTER]. When the calculator is set in REAL mode, we get an error message indicating that the answer is nonreal.

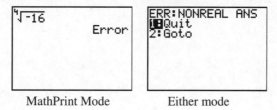

MathPrint Mode   Either mode

RATIONAL EXPONENTS

We can add and subtract rational exponents on a graphing calculator.

Section R.7, Example 9 (a) Simplify and then, if appropriate, write radical notation for $x^{5/6} \cdot x^{2/3}$.

To find this product we first add the exponents, expressing the sum as a fraction. Press [5] [÷] [6] [+] [2] [÷] [3] [MATH] [1] [ENTER]. These keystrokes tell the calculator to add $\frac{5}{6}$ and $\frac{2}{3}$. Then they access the MATH submenu of the MATH menu, copy item 1 "▶ Frac" to the home screen, and display the result as a fraction.

Alternatively, we can use the the n/d fraction template in the [F1] shortcut menu. Press [ALPHA] [F1] [ENTER] (or [1] [ENTER]) [5] [▶] [6] [▶] [+] [ALPHA] [F1] [ENTER] (or [1] [ENTER]) [2] [▶] [3] [▶] [ENTER].

```
5/6+2/3▶Frac
                    3
                    ─
                    2
```

MathPrint Mode

```
5   2
─ + ─
6   3
                    3
                    ─
                    2
```

MathPrint Mode using
the n/d fraction template
in the F1 shortcut menu.

Then, proceeding as on pages 49–50 of the text, we find that the final result in radical notation is $x\sqrt{x}$.

Chapter 1 Graphs, Functions, and Models

GRAPHING EQUATIONS

Section 1.1, Example 4 Graph $3x - 5y = -10$.

First we solve for y as shown in the text, obtaining $y = \frac{3}{5}x + 2$. Then press $\boxed{Y=}$ to access the equation-editor screen. Clear any equations that are present. To clear an entry for Y_1, for example, position the cursor beside "$Y_1 =$" and press $\boxed{CLEAR}$. Do this for each existing entry. Also turn off any plots that are turned on. If the name of a plot is highlighted at the top of the equation-editor screen, it is turned on. To turn it off, position the cursor over it and press $\boxed{ENTER}$. The name will no longer be highlighted, indicating that it is now turned off.

Next enter the equation by positioning the cursor beside "$Y_1 =$" and pressing $\boxed{(}$ $\boxed{3}$ $\boxed{\div}$ $\boxed{5}$ $\boxed{)}$ $\boxed{X,T,\Theta,n}$ $\boxed{+}$ $\boxed{2}$. Although the parentheses are not necessary, the equation is more easily read when they are used.

```
Plot1 Plot2 Plot3
\Y1B(3/5)X+2
\Y2=
\Y3=
\Y4=
\Y5=
\Y6=
\Y7=
```

Next we set the viewing window. This is the portion of the coordinate plane that appears on the calculator's screen. It is defined by the minimum and maximum values of x and y: Xmin, Xmax, Ymin, and Ymax. The notation [Xmin, Xmax, Ymin, Ymax] is used to represent these window settings or dimensions. For example, [−12, 12, −8, 8] denotes a window that displays the portion of the x-axis from −12 to 12 and the portion of the y-axis from −8 to 8. In addition, the distance between tick marks on the axes is defined by the settings Xscl and Yscl. In this manual Xscl and Yscl will be assumed to be 1 unless noted otherwise. The Xres setting sets the pixel resolution and ΔX determines the increments between X-values when the graph is being traced. We leave these values at their default settings of 1 and 0.2127659574…, respectively. The window corresponding to the settings [−20, 30, −12, 20], Xscl = 5, Yscl = 2, Xres = 1, ΔX = 0.2127659574… is shown on the next page.

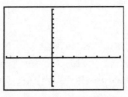

9

Press the WINDOW key on the top row of the keypad to display the current window settings on your graphing calculator. The standard settings are shown below.

```
WINDOW
 Xmin=-10
 Xmax=10
 Xscl=1
 Ymin=-10
 Ymax=10
 Yscl=1
↓Xres=1
```

To change a setting, position the cursor beside the setting you wish to change and enter the new value. For example, to change from the standard settings to [−20, 30, −12, 20], Xscl = 5, Yscl = 2, on the WINDOW screen, position the cursor beside Xmin, then press (-) 2 0 ENTER 3 0 ENTER 5 ENTER (-) 1 2 ENTER 2 0 ENTER 2 ENTER. Recall that the (-) key in the bottom row of the keypad must be used to enter a negative number. The − key in the right column of the keypad is used for the subtraction operation. The ▼ key may be used instead of ENTER after typing each window setting. To see this window, press the GRAPH key on the top row of the keypad. This is the window shown above.

QUICK TIP: To return quickly to the standard window setting [−10, 10,−10, 10], Xscl = 1, Yscl = 1, press ZOOM 6.

The standard window is a good choice for the graph of the equation $y = \frac{3}{5}x + 2$. Either enter these dimensions in the WINDOW screen and then press GRAPH to see the graph or simply press ZOOM 6 to select the standard window and see the graph.

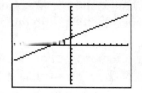

THE TABLE FEATURE

A table of x- and y-values representing ordered pairs that are solutions of an equation can be displayed on a graphing calculator.

Section 1.1, Example 5 Create a table of ordered pairs that are solutions of the equation $y = x^2 - 9x - 12$.

First we enter the equation on the equation-editor screen. Then press 2nd [TBLSET] to display the table set-up screen. (TblSet is the second function associated with the WINDOW key.) You can choose to supply the x-values yourself or you can set the graphing calculator to supply them. To have the graphing calculator supply the x-values, set

"Indpnt" to "Auto" by positioning the cursor over "Auto" and pressing [ENTER]. "Depend" should also be set to "Auto."

When "Indpnt" is set to "Auto," the graphing calculator will supply values for x, beginning with the value specified as TblStart and continuing by adding the value of ΔTbl to the preceding value for x. We will display a table of values that starts with $x = -3$ and adds 1 to the preceding x-value. Press [(-)] [3] [▼] [1] or [(-)] [3] [ENTER] [1] to select a minimum x-value of -3 and an increment of 1. To display the table press [2nd] [TABLE]. (TABLE is the second operation associated with the [GRAPH] key.) We can use the [▲] and [▼] keys to scroll up and down in the table to find values other than those shown here.

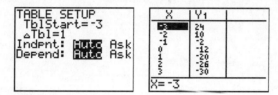

GRAPHING CIRCLES

If the center and radius of a circle are known, the circle can be graphed using the Circle feature from the DRAW menu.

Section 1.1, Example 12 Graph $(x-2)^2 + (y+1)^2 = 16$.

The center of this circle is $(2, -1)$ and its radius is 4. To graph it using the Circle feature from the DRAW menu first press [Y=] and clear all previously entered equations. Then select a square window. (See page 68 of the text for a discussion on squaring the viewing window.) We will use [−9, 9, −6, 6]. Press [2nd] [QUIT] to go to the home screen. Then press [2nd] [DRAW] [9] to display "Circle(." (DRAW is the second operation associated with the [PRGM] key.) Enter the coordinates of the center and the radius, separating the entries by commas, and close the parentheses: [2] [,] [(-)] [1] [,] [4] [)] [ENTER].

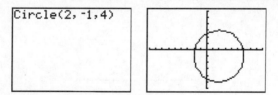

To clear this circle from the Graph screen, use the ClrDraw option from the DRAW menu. Press [2nd] [DRAW] [ENTER] or [2nd] [DRAW] [1] to do this.

FINDING FUNCTION VALUES

When a formula for a function is given, function values can be found in several ways.

Section 1.2, Example 4 (b) For $f(x) = 2x^2 - x + 3$, find $f(-7)$.

Method 1: Substitute the inputs directly in the formula. Press 2 $($ $(-)$ 7 $)$ x^2 $-$ $($ $(-)$ 7 $)$ $+$ 3 $\boxed{\text{ENTER}}$. Although it is not necessary to use the second set of parentheses, they allow the expression to be read more easily so we include them here.

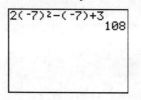

Method 2: Enter $Y_1 = 2x^2 - x + 3$ on the "Y =" screen. Then press $\boxed{\text{2nd}}$ $[\text{QUIT}]$ to go to the home screen. To find $f(-7)$, the value of Y_1 when $x = -7$, press $(-)$ 7 $\boxed{\text{STO►}}$ $\boxed{\text{X,T,}\Theta,n}$ $\boxed{\text{ALPHA}}$ $[:]$ $\boxed{\text{VARS}}$ $\boxed{►}$ 1 1 $\boxed{\text{ENTER}}$. ($[:]$ is the ALPHA operation associated with the $\boxed{\cdot}$ key.) This series of keystrokes stores -7 as the value of x and then substitutes it in the function Y_1.

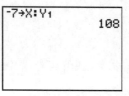

Method 3: Enter $Y_1 = 2x^2 - x + 3$ on the "Y =" screen and press $\boxed{\text{2nd}}$ $[\text{QUIT}]$ to go to the home screen. To find $f(-7)$, press $\boxed{\text{VARS}}$ $\boxed{►}$ 1 1 $($ $(-)$ 7 $)$ $\boxed{\text{ENTER}}$. Note that this entry closely resembles function notation.

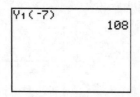

Method 4: The TABLE feature can also be used to find function values. Enter $Y_1 = 2x^2 - x + 3$ on the "Y =" screen. Then set up a table in "Ask" mode, by pressing $\boxed{\text{2nd}}$ $[\text{TBLSET}]$, moving the cursor over "Ask" on the Indpnt: line and pressing $\boxed{\text{ENTER}}$. In "Ask" mode, you supply x-values and the calculator returns the corresponding y-values.

The settings for TblStart and ΔTbl are irrelevant in this mode. Press 2nd [TABLE] to display the TABLE screen. Then press (-) 7 [ENTER] to find $f(-7)$.

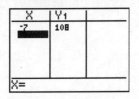

Method 5: We can also use the Value feature from the CALC menu to find $f(-7)$. To do this, graph $Y_1 = 2x^2 - x + 3$ in a window that includes the x-value -7. We will use the standard window. Then press 2nd [CALC] 1 or 2nd [CALC] [ENTER] to access the CALC menu and select item 1, Value. Now supply the desired x-value by pressing (-) 7. Press [ENTER] to see X = -7, Y = 108 at the bottom of the screen, Thus, $f(-7) = 108$.

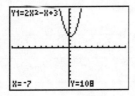

GRAPHS OF FUNCTIONS

Three functions are graphed in **Section 1.2, Example 5**. The TI-84 Plus graphing calculator does not use function notation. Consequently, we must first replace function notation with y when we graph a function on this calculator. For example, to graph $f(x) = x^2 - 5$ replace $f(x)$ with y. Then enter the equation $y = x^2 - 5$ on the equation-editor screen and graph it as described on pages 9–10 of this manual.

LINEAR REGRESSION

We can use the Linear Regression feature in the STAT CALC menu to fit a linear equation to a set of data.

Section 1.4, Example 7 The following table shows the U.S. Gross Domestic Product, where the year 1980 is represented by $x = 0$. Fit a regression line to the data. Then use the function to estimate the GDP in 2014.

Year	Number of Years since 1980, x	Gross Domestic Product (GDP) (in trillions)
1980	0	$ 2.8
1985	5	4.2
1990	10	5.8
1995	15	7.4
2000	20	10.0
2005	25	12.6
2010	30	14.7

We will enter the data as ordered pairs on the STAT list editor screen. To clear any existing lists press [STAT] [4] [2nd] [L1] [,] [2nd] [L2] [,] [2nd] [L3] [,] [2nd] [L4] [,] [2nd] [L5] [,] [2nd] [L6] [ENTER]. ([L1] through [L6] are the second operations associated with the numeric keys 1 through 6.) The lists can also be cleared by first accessing the STAT list editor screen by pressing [STAT] [ENTER] or [STAT] [1]. These keystrokes display the STAT EDIT menu and then select the Edit option from that menu. Then, for each list that contains entries, use the arrow keys to move the cursor to highlight the name of the list at the top of the column and then press [CLEAR] [▾] or [CLEAR] [ENTER].

Once the lists are cleared, we can enter the coordinates of the points. If the lists were cleared using ClrList, press [STAT] [1] to get to the list entry screen. We will enter the first coordinates in L_1 as the number of years since 1980 and the second coordinates, the GDP, in L_2. Position the cursor at the top of column L_1, below the L_1 heading. To enter 0, press [0] [ENTER]. Continue typing the first coordinates, 5, 10, 15, 20, 25, and 30 in order, each followed by [ENTER]. The entries can be followed by [▾] rather than [ENTER] if desired. Press [▸] to move to the top of column L_2. Type the second coordinates, 2.8, 4.2, 5.8, 7.4, 10.0, 12.6, and 14.7, in succession, each followed by [ENTER] or [▾]. Note that the coordinates of each point must be in the same position in both lists.

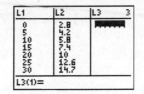

The graphing calculator can then plot these points, creating a scatterplot of the data. To do this we turn on the STAT PLOT feature. To access the STAT PLOT screen, press 2nd [STAT PLOT]. (STAT PLOT is the second operation associated with the Y= key in the upper left-hand corner of the keypad.) We will use Plot1. Access it by highlighting 1 and pressing ENTER or simply by pressing 1. Position the cursor over On and press ENTER to turn on Plot1. Now we select Type, Xlist, Ylist, and Mark. Type determines the style of the plot. We will select the first type shown, a scatterplot. Xlist and Ylist designate the STAT lists that supply the first and second coordinates, respectively, of the points in the plot. The last item, Mark, allows us to choose a box, a cross, or a dot for each point. Here we have selected a box. To select Type and Mark, position the cursor over the appropriate selection and press ENTER. Use the [L1] and [L2] keys (associated with the 1 and 2 numeric keys) to select L_1 and L_2 as Xlist and Ylist, respectively. The entries should be as shown below.

If the plot has previously been set up as desired, you do not need to follow the steps above. Instead, you can turn the plot on from the equation-editor, or "Y =", screen. To use this alternative, press Y= to go to this screen. Then, assuming Plot1 has not yet been turned on, position the cursor over Plot1 and press ENTER. Plot1 will now be highlighted.

Note that there should be no equations entered on the equation-editor screen. If there are equations present, clear them as described on page 9 of this manual. If this is not done, the equations that are currently entered will be graphed along with the data points in the scatterplot.

Now we select a window. The x-values range from 0 through 30 and the y-values from 2.8 through 14.7, so one good choice is [−5, 35, 0, 16], Xscl = 5, Yscl = 2. We can enter these dimensions and then press GRAPH to see the scatterplot. Instead of entering the window dimensions directly, we can press ZOOM 9 after entering the coordinates of the points in lists, turning on Plot1, and selecting Type, Xlist, Ylist, and Mark. This activates the ZoomStat operation which automatically defines a viewing window that displays all the points. We use the window [−5, 35, 0, 16], Xscl = 5, Yscl = 2 to produce the scatterplot shown below.

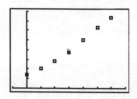

The calculator's linear regression feature can be used to fit a linear equation to the data. Once the data have been entered in the lists, press STAT ▶ 4 to select LinReg($ax + b$) from the STAT CALC menu. On a TI-84 Plus with MathPrint installed, the STAT WIZARD window will open. The wizard will prompt for required and optional arguments for the LinReg($ax + b$) function. The x-values are stored in L_1 and the y-values are stored in L_2. It is not necessary to include a value for FreqList. Use the next entry, Store RegEQ: to indicate that the linear regression equation should be stored as Y_1 by pressing VARS ▶ 1 1. Alternatively, we can use the shortcut menu by pressing ALPHA [F4] ENTER. Refer to page *ii* in the preface for more information about the shortcut menu.

YVARShortcut menu

To display the coefficients a and b of the regression equation $y = ax + b$, press ENTER.

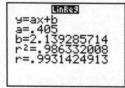

If the diagnostics have been turned on, values for r^2 and r will be displayed. These numbers indicate how well the regression line fits the data. See the discussion of the coefficient of linear correlation, r, on page 116 of the text.

On a calculator with MathPrint, both the STAT WIZARD and STAT DIAGNOSTICS can be turned on or off using the MODE screen. See page 1 of this manual. Alternatively, DiagnosticOn mode can be selected from the CATALOG menu by pressing [2nd] [CATALOG] and use [▼] to position the triangular selection cursor beside DiagnosticOn. To alleviate the tedium of scrolling through many items to reach DiagnosticOn, press [D] after pressing [2nd] [CATALOG] to move quickly to the first catalog item that begins with the letter [D]. ([D] is the ALPHA operation associated with the [x⁻¹] key.) Note that it is not necessary to press [ALPHA] before [D] when the catalog is displayed. Then use [▼] to scroll to DiagnosticOn. Press [ENTER] to paste this instruction to the home screen and then press [ENTER] a second time to set the mode. To select DiagnosticOff mode, press [2nd] [CATALOG], position the selection cursor beside DiagnosticOff, press [ENTER] to paste this instruction to the home screen, and then press [ENTER] again to set this mode. Similarly, it is also possible to turn the STAT WIZARD on or off using the CATALOG menu.

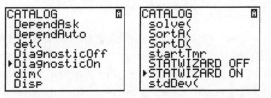

If you have not indicated that the equation should be saved as Y_1, this can be done immediately after the regression equation is found. (If you selected DiagnosticOn *after* finding the regression equation, you will have to find the equation again in order to copy it to the equation-editor screen.) To copy the regression equation as Y_1, for example, first note that any previous entry in Y_1 should be cleared. Press [Y=] and position the cursor beside Y_1. Then press [VARS] [5] [▶] [▶] [1]. These keystrokes select Statistics from the VARS menu, then select the EQ (Equation) submenu, and finally select the RegEq (Regression Equation) from this submenu.

```
Plot1 Plot2 Plot3
\Y1 ◼.405X+2.1392
857142857
\Y2=
\Y3=
\Y4=
\Y5=
\Y6=
```
Classic Mode

It is possible to select a *y*-variable to which it will be stored on the equation editor screen before the regression equation is found. On a calculator with MathPrint, the STAT WIZARD must be turned off. After the data have been stored in the lists and the equation previously entered as Y_1 has been cleared, press [STAT] [▶] [4] [VARS] [▶] [1] [1] [ENTER]. The coefficients of the regression equation will be displayed on the home screen, and the regression equation will also be stored as Y_1 on the equation-editor screen.

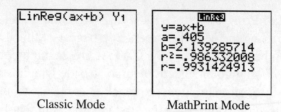

Classic Mode   MathPrint Mode

Once the regression equation has been copied to the equation-editor screen, we can see the graph of the equation along with the scatterplot by pressing $\boxed{\text{GRAPH}}$.

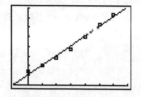

To estimate the GDP in 2014, evaluate the regression equation for $x = 34$. (2014 is 34 years after 1980.) Use any of the methods for evaluating a function presented earlier in this chapter. (See pages 12–13 of this manual.) We will use function notation on the home screen. When $x = 34$, $y \approx 15.9$, so we estimate that the gross domestic product will be about \$15.9 trillion in 2014.

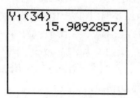

Turn off the STAT PLOT, either using the STAT PLOT screen or the "Y=" screen, before graphing other functions.

THE INTERSECT METHOD

We can use the Intersect feature from the CALC menu to solve equations. We call this the **Intersect method**.

Section 1.5, Example 1 Solve $\dfrac{3}{4}x - 1 = \dfrac{7}{5}$.

Press $\boxed{\text{Y=}}$ to go to the equation-editor screen. Clear any existing entries and then enter $Y_1 = \frac{3}{4}x - 1$ and $Y_2 = \frac{7}{5}$. The solution of the original equation is the first coordinate of the point of intersection of the graphs of Y_1 and Y_2. Graph the equations in a window that displays the point of intersection. The window [−5, 5, −5, 5] is a good choice.

Next we press $\boxed{\text{2nd}}$ $\boxed{\text{CALC}}$ $\boxed{5}$ to select the Intersect feature from the CALC menu. (CALC is the second operation associated with the $\boxed{\text{TRACE}}$ key.) The query "First curve?" appears at the bottom of the screen. The blinking cursor is positioned on the graph of Y_1. This is indicated by the notation $Y_1 = (3/4)x - 1$ in the upper left-hand corner of the screen. Press $\boxed{\text{ENTER}}$ to indicate that this is the first curve involved in the intersection. Next the query "Second curve?" appears at the bottom of the screen. The blinking cursor is now positioned on the graph of Y_2 and the notation $Y_2 = 7/5$ should appear in the top left-hand corner of the screen. Press $\boxed{\text{ENTER}}$ to indicate that this is the second curve. We identify the curves for the calculator since we could have as many as ten graphs on the screen at once. After we identify the second curve, the query "Guess?" appears at the bottom of the screen. Use the right and left arrow keys to move the blinking cursor close to the point of intersection of the graphs or key in a number that appears to be close to the x-coordinate of this point. This provides the calculator with a guess as to the coordinates of this point. We do this since some pairs of curves can have more than one point of intersection. When the cursor is positioned or the x-value is entered, press $\boxed{\text{ENTER}}$ a third time. Now the coordinates of the point of intersection appear at the bottom of the screen.

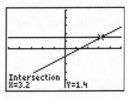

We see that the first coordinate of the point of intersection is 3.2. We can find fraction notation for the solution by using the ▶Frac feature from the MATH menu. To do this first press $\boxed{\text{2nd}}$ $\boxed{\text{QUIT}}$ to go to the home screen. The x- and y-coordinates of the point of intersection are stored in the calculator as X and Y, respectively. To convert the decimal notation for X to a fraction, press $\boxed{\text{X,T,Θ,n}}$ $\boxed{\text{MATH}}$ $\boxed{1}$ $\boxed{\text{ENTER}}$. These keystrokes tell the calculator to use X, and then they access the MATH submenu of the MATH menu, copy item 1 "▶Frac" to the home screen, and display the conversion. We see that the first coordinate of the point of intersection is $\frac{16}{5}$. The solution of the equation is 3.2, or $\frac{16}{5}$.

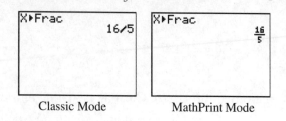

Classic Mode   MathPrint Mode

THE ZERO METHOD

The Zero feature from the CALCULATE menu can be used to find the zeros of a function or to solve an equation in the form $f(x) = 0$. We call this the **Zero method**.

Section 1.5, Example 11 Find the zero of $f(x) = 5x - 9$.

On the equation-editor screen, clear any existing entries and then enter $Y_1 = 5x - 9$. Now graph the function in a viewing window that shows the x-intercept clearly. The standard window is a good choice.

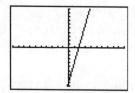

Press 2nd [CALC] to display the CALC menu. Then press 2 to select the Zero feature. We are first prompted to select a left bound for the zero. This means that we must choose an x-value that is to the left of the x-intercept. This can be done by using the left- and right-arrow keys to move the cursor to a point on the curve to the left of the x-intercept or by keying in a value less than the first coordinate of the intercept.

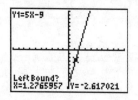

Once this is done press ⌐ENTER⌐. Now we are prompted to select a right bound that is to the right of the *x*-intercept. Again, this can be done by using the arrow keys to move the cursor to a point on the curve to the right of the *x*-intercept or by keying in a value greater than the first coordinate of the intercept.

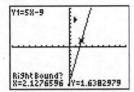

Press ⌐ENTER⌐ again. Finally we are prompted to make a guess as to the value of the zero. Move the cursor to a point close to the zero or key in a value.

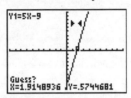

Finally, press ⌐ENTER⌐ a third time. We see that $y = 0$ when $x = 1.8$, so 1.8 is the zero of the function.

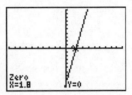

If a function has more than one zero, the Zero feature can be used as many times as necessary to find all of them.

Chapter 2 More on Functions

THE MAXIMUM AND MINIMUM FEATURES

Section 2.1, Example 2 Use a graphing calculator to determine any relative maxima or minima of the function $f(x) = 0.1x^3 - 0.6x^2 - 0.1x + 2$.

First graph $Y_1 = 0.1x^3 - 0.6x^2 - 0.1x + 2$ in a window that displays the relative extrema of the function. Trial and error reveals that one good choice is [−4, 6, −3, 3]. Observe that a relative maximum occurs near $x = 0$ and a relative minimum occurs near $x = 4$.

To find the relative maximum, first press [2nd] [CALC] [4] or [2nd] [CALC] [▼] [▼] [▼] [ENTER] to select the Maximum feature from the CALCULATE menu. We are prompted to select a left bound for the relative maximum. This means that we must choose an x-value that is to the left of the x-value of the point where the relative maximum occurs. This can be done by using the left- and right-arrow keys to move the cursor to a point to the left of the relative maximum or by keying in an appropriate value.

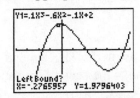

Once this is done, press [ENTER]. Now we are prompted to select a right bound. We move the cursor to a point to the right of the relative maximum or we can key in an appropriate value.

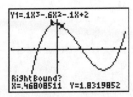

Press [ENTER] again. Finally we are prompted to guess the x-value at which the relative maximum occurs. Move the cursor close to the relative maximum point or key in an x-value.

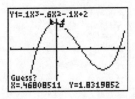

Press ENTER a third time. We see that a relative maximum function value of approximately 2.004 occurs when $x \approx -0.082$.

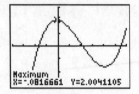

To find the relative minimum, select the Minimum feature from the CALCULATE menu by pressing 2nd [CALC] 3 or 2nd [CALC] ▼ ▼ ENTER. Select left and right bounds for the relative minimum and guess the x-value at which it occurs as described above. We see that a relative minimum function value of approximately -1.604 occurs when $x \approx 4.082$.

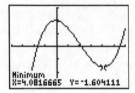

PIECEWISE FUNCTIONS

We can graph piecewise functions using the TEST menu operations. Press 2nd [TEST] to display this menu. (The TEST menu is the second operation on the MATH key.)

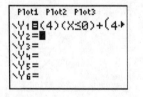

Section 2.1, Example 7 Graph the function defined as

$$f(x) = \begin{cases} 4, & \text{for } x \le 0, \\ 4 - x^2, & \text{for } 0 < x \le 2, \\ 2x - 6, & \text{for } x > 2. \end{cases}$$

If we use MathPrint mode, the equation will scroll off the screen.

So, we will use Classic mode instead. Press $\boxed{\text{MODE}}$ and then select FUNC mode and CLASSIC mode. Press $\boxed{\text{2nd}}$ [QUIT] to return to the home screen.

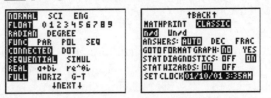

Turn off or clear all functions and stat plots in the "Y=" screen. (See pages 9 and 15 of this manual.) Position the cursor beside "Y_1=" and then key the function as follows: $\boxed{(}$ $\boxed{(}$ $\boxed{4}$ $\boxed{)}$ $\boxed{(}$ $\boxed{X,T,\Theta,n}$ $\boxed{\text{2nd}}$ [TEST] $\boxed{6}$ $\boxed{0}$ $\boxed{)}$ $\boxed{+}$ $\boxed{(}$ $\boxed{4}$ $\boxed{-}$ $\boxed{X,T,\Theta,n}$ $\boxed{x^2}$ $\boxed{)}$ $\boxed{(}$ $\boxed{0}$ $\boxed{\text{2nd}}$ [TEST] $\boxed{5}$ $\boxed{X,T,\Theta,n}$ $\boxed{)}$ $\boxed{(}$ $\boxed{X,T,\Theta,n}$ $\boxed{\text{2nd}}$ [TEST] $\boxed{6}$ $\boxed{2}$ $\boxed{)}$ $\boxed{+}$ $\boxed{(}$ $\boxed{2}$ $\boxed{X,T,\Theta,n}$ $\boxed{-}$ $\boxed{6}$ $\boxed{)}$ $\boxed{(}$ $\boxed{X,T,\Theta,n}$ $\boxed{\text{2nd}}$ [TEST] $\boxed{3}$ $\boxed{2}$ $\boxed{)}$.

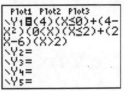

This function is entered with each piece entered in parentheses followed by the x-interval, also entered in parentheses. The pieces of the function are separated with "+" signs. Note that the compound inequality $0 < x \le 2$ is entered in two pieces, as $(0 < x)(x \le 2)$. The graph of the function is shown below, using the window $[-6, 6, -6, 6]$.

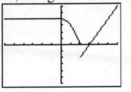

THE GREATEST INTEGER FUNCTION

The greatest integer function is found in the MATH NUM menu and is denoted "int." To find int(1.9) press $\boxed{\text{MATH}}$ $\boxed{\blacktriangleright}$ $\boxed{5}$ $\boxed{1}$ $\boxed{.}$ $\boxed{9}$ $\boxed{)}$ $\boxed{\text{ENTER}}$. Note that the calculator supplies a left parenthesis and we close the parentheses after entering 1.9.

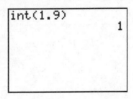

We can also graph the greatest integer function.

Section 2.1, Example 9 Graph $f(x) = [\![x]\!]$.

Clear any functions that have previously been entered on the equation-editor screen as discussed on page 9 in this manual. Then position the cursor beside "$Y_1 =$" and select the greatest integer function from the MATH NUM menu by pressing MATH ▶ 5 X,T,Θ,n) . Select a window and press GRAPH . The window [−6, 6, −6, 6] is shown here.

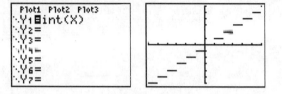

THE COMPOSITION OF FUNCTIONS

We can evaluate composite functions on a graphing calculator.

Section 2.3, Example 1 (b) Given that $f(x) = 2x - 5$ and $g(x) = x^2 - 3x + 8$, find $(f \circ g)(7)$ and $(g \circ f)(7)$.

On the equation-editor screen enter $Y_1 = 2x - 5$ and $Y_2 = x^2 - 3x + 8$. Then $(f \circ g)(7) = (Y_1 \circ Y_2)(7)$, or $Y_1(Y_2(7))$, and $(g \circ f)(7) = (Y_2 \circ Y_1)(7)$, or $Y_2(Y_1(7))$. To find these function values press 2nd [QUIT] to go to the home screen. Then enter $Y_1(Y_2(7))$ by pressing VARS ▶ 1 1 (VARS ▶ 1 2 (7))) ENTER . Enter $Y_2(Y_1(7))$ by pressing VARS ▶ 1 2 (VARS ▶ 1 1 (7))) ENTER .

```
Y₁(Y₂(7))
                67
Y₂(Y₁(7))
                62
```

Chapter 3 Quadratic Functions and Equations; Inequalities

OPERATIONS WITH COMPLEX NUMBERS

Operations with complex numbers can be performed on the TI-84 Plus. First set the calculator in the complex $a + bi$ mode by pressing [MODE], positioning the cursor over $a + bi$, and pressing [ENTER].

Section 3.1, Example 2
(a) Add: $(8 + 6i) + (3 + 2i)$.

To find this sum go to the home screen and press [8] [+] [6] [2nd] [i] [+] [3] [+] [2] [2nd] [i] [ENTER]. (The number [i] is the second operation associated with the [.] key.) Note that it is not necessary to include parentheses when we are adding.

(b) Subtract: $(4 + 5i) - (6 - 3i)$

Press [4] [+] [5] [2nd] [i] [−] [(] [6] [−] [3] [2nd] [i] [)] [ENTER]. Note that the parentheses must be included as shown so that the entire number $6 - 3i$ is subtracted.

```
8+6i+3+2i
                11+8i
4+5i-(6-3i)
                -2+8i
```

Section 3.1, Example 3

(a) Multiply: $\sqrt{-16} \cdot \sqrt{-25}$.

The keystrokes are slightly different in each mode. In Classic mode, Press [2nd] [√] [(-)] [1] [6] [)] [2nd] [√] [(-)] [2] [5] [)] [ENTER]. In MathPrint mode, Press [2nd] [√] [(-)] [1] [6] [▶] [2nd] [√] [(-)] [2] [5] [ENTER].

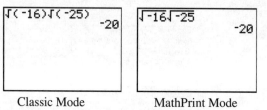

Classic Mode   MathPrint Mode

(b) Multiply: $(1 + 2i)(1 + 3i)$.

The keystrokes are the same in each mode. Press [(] [1] [+] [2] [2nd] [i] [)] [(] [1] [+] [3] [2nd] [i] [)] [ENTER].

(c) Multiply: $(3 - 7i)^2$.

The keystrokes are the same in each mode. Press [(] [3] [−] [7] [2nd] [i] [)] [x²] [ENTER].

```
(1+2i)(1+3i)
              -5+5i
(3-7i)²
           -40-42i
```

Section 3.1, Example 6 Divide $2 - 5i$ by $1 - 6i$.

We have $\dfrac{2-5i}{1-6i}$, or $(2-5i) \div (1-6i)$. Press ⎡(⎤ ⎡2⎤ ⎡−⎤ ⎡5⎤ ⎡2nd⎤ $[i]$ ⎡)⎤ ⎡÷⎤ ⎡(⎤ ⎡1⎤ ⎡−⎤ ⎡6⎤ ⎡2nd⎤

$[i]$ ⎡)⎤. To express a and b in fraction form in the quotient $a + bi$, we also select the ▶Frac

feature from the MATH menu by pressing ⎡MATH⎤ ⎡ENTER⎤ or ⎡MATH⎤ ⎡1⎤. Finally, press ⎡ENTER⎤
to see the quotient.

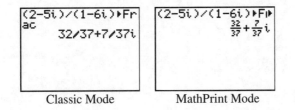

Classic Mode   MathPrint Mode

Chapter 4 Polynomial Functions and Rational Functions

POLYNOMIAL MODELS

We can fit polynomial functions to data on a graphing calculator.

Section 4.1, Example 9 The graph below shows adjusted vehicle fuel efficiency in miles per gallon for various years from 2000–2010 along with future corporate average fuel economy (CAFE) standards.

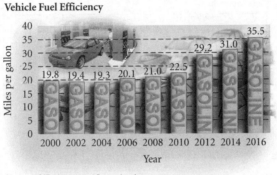

Vehicle Fuel Efficiency

Source: U.S. Environmental Protection Agency

(a) Model the data with a quadratic function, a cubic function and a quartic function. Let the first coordinate of each data point be the number of years after 2000; that is, enter the data as (0, 19.8), (2, 19.4), (4, 19.3), and so on. Then using R^2, the coefficient of determination, decide which function is the best fit.

Turn on STAT DIAGNOSTICS and STAT WIZARD as described on page 17 of this manual. Then, enter the data in lists as described on pages 14–15 of this manual.

To model the data with a quadratic function, select quadratic regression from the STAT CALC menu by pressing $\boxed{\text{STAT}}$ $\boxed{\blacktriangleright}$ $\boxed{5}$. The STAT WIZARD will appear. If the Xlist is not given as L_1, then place the cursor next to Xlist: and press $\boxed{\text{2nd}}$ $[L1]$. Similarly, make sure that Ylist is given as L_2. The entry for FreqList should be blank. Store the regression equation as Y_1 by placing the cursor next to Store RegEQ: and pressing either $\boxed{\text{ALPHA}}$ $[F4]$ $\boxed{\text{ENTER}}$ (to access the YVARS shortcut menu) or $\boxed{\text{VARS}}$ $\boxed{\blacktriangleright}$ $\boxed{1}$ $\boxed{1}$. Finally, press $\boxed{\text{ENTER}}$ twice. The graphing calculator displays the coefficients of a quadratic function $y = ax^2 + bx + c$. The function is shown on the next page.

To model the data with a cubic function, select cubic regression from the STAT CALC menu by pressing [STAT] [▶] [6], then follow the same process as for the quadratic regression. Store the regression equation as Y_2 by placing the cursor next to Store RegEQ: and pressing either [ALPHA] [F4] [▼] [ENTER] (to access the YVARS shortcut menu) or [VARS] [▶] [1] [2]. Finally, press [ENTER] twice. The graphing calculator displays the coefficients of a cubic function $y = ax^3 + bx^2 + cx + d$.

In a similar manner, find a quartic function. This is choice [7] in the STAT CALC menu. Store the regression equation as Y_3. The graphing calculator displays the coefficients of a quartic function $y = ax^4 + bx^3 + cx^2 + dx + e$.

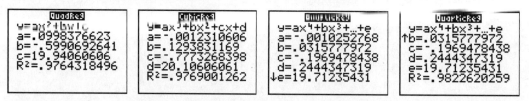

Since the R^2-value for the quartic function is the closest R^2-value to 1, the quartic function is the best fit.

$$f(x) = -0.0010252768x^4 + 0.0315777972x^3 - 0.1969478438x^2$$
$$+ 0.2444347319x + 19.71235431.$$

It is possible to find a regression without using the STAT WIZARD. For example, to find a quadratic regression with STAT WIZARD turned off, start as above, by pressing [STAT] [▶] [5]. This pastes QuadReg to the home screen. Then press [2nd] [L1] [,] [2nd] [L2] [ENTER]. See the left figure below. In order to paste the regression equation to Y_1, press [Y=] and clear any existing equation from Y_1 as described on page 9 of this manual. With the cursor positioned next to Y_1=, press [VARS] [5] [▶] [▶] [1].

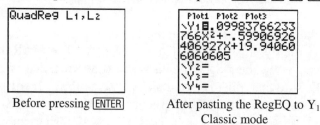

Before pressing [ENTER]   After pasting the RegEQ to Y_1
                                            Classic mode

The QuadReg screen shown above will appear after pressing [ENTER]. Use a similar procedure to find the cubic regression equation and the quartic regression equation.

(b) Graph the function with the best fit, along with the scatterplot of the data.

A scatterplot of the data can be graphed as described on page 16 of this manual. Turn off the graph for Y_1 by placing the cursor on the equal sign ("=") and pressing [ENTER]. Similarly, turn off the graph for Y_2.

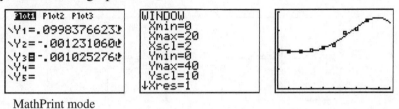

MathPrint mode

(c) Use the answer to part (a) to estimate the number of miles per gallon in 2005, in 2011, in 2017, and in 2021.

The function can be evaluated using one of the methods on pages 12 and 13. We will set up a table in "Ask" mode, using the X-values 5 (for 2005), 11 (for 2011), 17 (for 2017), and 21 (for 2021).

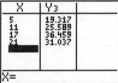

Thus, we estimate the fuel efficiency to be 19.3 mpg in 2005, 25.6 mpg in 2011, 36.5 mpg in 2017, and 31.0 mpg in 2021.

GRAPHING RATIONAL FUNCTIONS

Section 4.5, Example 10 Graph: $g(x) = \dfrac{x-2}{x^2-x-2}$.

As explained in the text, the graph of $g(x)$ is the graph of

$$y = \frac{1}{x+1}$$

with the point $\left(2, \frac{1}{3}\right)$ missing. To use a graphing calculator to produce the graph with a "hole" at $\left(2, \frac{1}{3}\right)$, use the ZDecimal feature from the ZOOM menu.

After entering

$$Y_1 = \frac{x-2}{x^2 - x - 2}$$

on the Y = screen, press ZOOM 4 to select this window and display the graph, shown on the right.

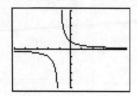

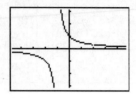

This graph is plotted in the window that is [−5, 5, −5, 5]. Notice that there is no "hole".

This graph is plotted using ZDecimal. Notice the "hole" at $\left(2, \frac{1}{3}\right)$.

To confirm that there is no point on the graph with a first coordinate of 2, try to use the Value feature from the CALC menu to find the value of the function when $x = 2$. We see that there is no y-value that corresponds to $x = 2$.

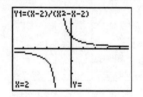

Chapter 5 Exponential Functions and Logarithmic Functions

GRAPHING AN INVERSE FUNCTION

The DrawInv operation can be used to graph a function and its inverse on the same screen. A formula for the inverse function need not be found in order to do this. The calculator must be set in FUNC mode when this operation is used.

Section 5.1, Example 7 Graph $f(x) = 2x - 3$ and $f^{-1}(x)$ using the same set of axes.

Enter $Y_1 = 2x - 3$, clear all other functions on the "Y =" screen, and select a window. Be sure that you are in CONNECTED mode. Go to the home screen and press [2nd] [DRAW] [8] to select the DrawInv operation. (DRAW is the second operation associated with the [PRGM] key.) Follow these keystrokes with [VARS] [▶] [1] [1] or [ALPHA] [F4] [ENTER] to select function Y_1. Press [ENTER] to see the graph of the function and its inverse. The graphs are shown here in the standard window.

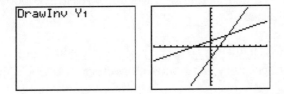

EVALUATING e^x, Log x, and Ln x

Use the calculator's scientific keys to evaluate e^x, log x, and ln x for specific values of x.

Section 5.2, Example 5 (a), (b) Find the values of e^3 and $e^{-0.23}$. Round to four decimal places.

To find e^3, press [2nd] [e^x] [3] [)] [ENTER] in Classic mode or [2nd] [e^x] [3] [ENTER] in MathPrint mode. (e^x is the second operation associated with the [LN] key.) The calculator returns 20.08553692. Thus, $e^3 \approx 20.0855$. The right parenthesis in Classic mode is not absolutely necessary, but it is included to make the expression appear complete.

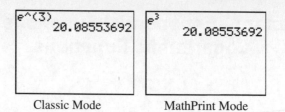

| Classic Mode | MathPrint Mode |

To find $e^{-0.23}$, press [2nd] [e^x] [(-)] [.] [2] [3] [)] [ENTER] in Classic mode or [2nd] [e^x] [(-)] [.] [2] [3] [ENTER] in MathPrint mode. Again, we enclose the exponent in parentheses in Classic mode so that the expression appears complete. The calculator returns .7945336025, so $e^{-0.23} \approx 0.7945$.

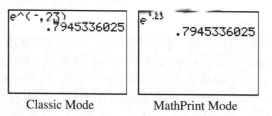

| Classic Mode | MathPrint Mode |

Section 5.3, Example 5 Find the values of log 645,778, log 0.0000239, and log (−3). Round to four decimal places.

To find log 645,778 press [LOG] [6] [4] [5] [7] [7] [8] [)] [ENTER] in either mode. Note that the final parenthesis is not necessary. The calculator returns 5.810083246. Thus, $\log 645,778 \approx 5.8101$.

To find log 0.0000239 press [LOG] [.] [0] [0] [0] [0] [2] [3] [9] [)] [ENTER] in either mode. Again, note that the final parenthesis is not necessary. The calculator returns −4.621602099, so $\log 0.0000239 \approx -4.6216$.

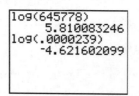

When the calculator is set in Real mode the keystrokes $\boxed{\text{LOG}}$ $\boxed{\text{(-)}}$ $\boxed{3}$ $\boxed{)}$ $\boxed{\text{ENTER}}$ produce the message ERR: NONREAL ANS, indicating that the result of this calculation is not a real number.

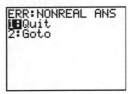

Section 5.3, Example 6 (a), (b), (c) Find the values of ln 645,778, ln 0.0000239, and ln (−5). Round to four decimal places.

To find ln 645,778 and ln 0.0000239 repeat the keystrokes used above to find log 645,778 and log 0.0000239 but press $\boxed{\text{LN}}$ rather than $\boxed{\text{LOG}}$. We find that ln 645,778 ≈ 13.3782 and ln 0.0000239 ≈ −10.6416.

```
ln(645778)
         13.37821107
ln(.0000239)
         -10.6416321
```

When the calculator is set in Real mode the keystrokes $\boxed{\text{LN}}$ $\boxed{\text{(-)}}$ $\boxed{5}$ $\boxed{)}$ $\boxed{\text{ENTER}}$ produce the message ERR: NONREAL ANS (shown at the top of this page), indicating that the result of this calculation is not a real number.

CHANGING LOGARITHMIC BASES

A TI-84 Plus calculator with MathPrint has a built-in operation called logBASE that will find a logarithm with any base. As discussed in the text, it is also possible to use the change-of-base formula,

$$\log_b M = \frac{\log_a M}{\log_a b},$$

where a and b are any logarithmic bases and M is any positive number to find a logarithm with a base other than 10 or e.

Section 5.3, Example 7 Find $\log_5 8$ using common logarithms.

The logBASE operation can be accessed from the MATH MATH menu or from the FUNC shortcut menu.

To use logBASE from the MATH MATH menu, with the calculator in MathPrint mode, press MATH ALPHA [A]. The logBASE template appears on the home screen. The cursor should appear in the base of the logarithm. Now press 5 ▶ 8 ▶ ENTER. The result is about 1.2920.

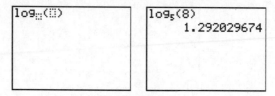

MathPrint Mode

To access the logBASE shortcut, with the calculator in MathPrint mode, press ALPHA [F2] [5]. The logBASE template shown above will appear on the home screen, then proceed as above.

It is also possible to use logBASE with the calculator in Classic mode. Press MATH ALPHA [A]. The Classic mode logBASE template appears on the home screen. Then press 8 , 5).

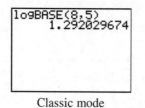

Classic mode

To use the change-of-base formula, let $a = 10$, $b = 5$, and $M = 8$, and substitute in the formula. We have

$$\log_5 8 = \frac{\log_{10} 8}{\log_{10} 5}.$$

To perform this computation press LOG 8) ÷ LOG 5) ENTER. Note that the parentheses must be closed in the numerator to enter the expression correctly. We also close the parentheses in the denominator for completeness. The result is about 1.2920. We could have let $a = e$ and then used natural logarithms to find $\log_5 8$ as well, as shown in Example 8 in the text.

```
log(8)/log(5)
          1.292029674
ln(8)/ln(5)
          1.292029674
```

Section 5.3, Example 9 Graph $y = \log_5 x$.

To graph the function on a graphing calculator, use either logBASE or the change-of-base formula. Here we use logBASE. Enter $Y_1 = \log_5 x$ on the $Y =$ screen as described earlier, select a window, and press $\boxed{\text{GRAPH}}$. The window shown is $[-1, 10, -3, 3]$.

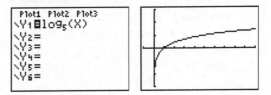

EXPONENTIAL AND LOGARITHMIC REGRESSION

In addition to the types of polynomial regression discussed earlier, exponential and logarithmic functions can be fit to data. The operations of entering data, making scatterplots, and graphing and evaluating these functions are the same as for linear regression functions. The procedures for copying a regression equation to the $Y =$ screen, graphing it, and using it to find function values are also the same. Note that the coefficient of correlation, r, will be displayed only if DiagnosticOn has been selected from the Mode screen or from the Catalog. (See page 17 in this manual.)

Section 5.6, Example 6 *Spending on Video Rentals.* The number of video stores is declining rapidly. Consumer spending on video rentals from subscription services, including Netflix, kiosks, and cable and satellite video-on-demand services has increased greatly in recent years, as shown in the table below.

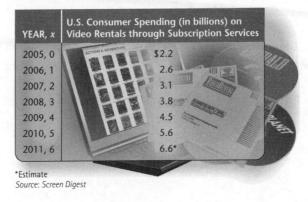

YEAR, x	U.S. Consumer Spending (in billions) on Video Rentals through Subscription Services
2005, 0	$2.2
2006, 1	2.6
2007, 2	3.1
2008, 3	3.8
2009, 4	4.5
2010, 5	5.6
2011, 6	6.6*

*Estimate
Source: Screen Digest

(a) Use a graphing calculator to fit an exponential function to the data.

Enter the data in lists as described on pages 14–15 of this manual. Then select exponential regression from the STAT CALC menu by pressing [STAT] [▸] [0]. On a TI-84 Plus with MathPrint installed and STAT WIZARD on, the STAT WIZARD window will open. The wizard will prompt for required and optional arguments for the ExpReg($ax + b$) function. Move the cursor through the list by pressing [▾] or [ENTER]. The x-values are stored in L_1 and the y-values are stored in L_2. It is not necessary to include a value for FreqList. Use the entry, Store RegEQ: to indicate that the exponential regression equation should be stored as Y_1 by pressing [VARS] [▸] [1] [1]. Alternatively, we can use the shortcut menu by pressing [ALPHA] [F4] [ENTER]. Press [ENTER] [ENTER], and the calculator displays the coefficient a and the base b for the exponential function $y = ab^x$.

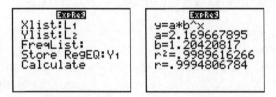

A scatterplot of the data can be graphed as described on pages 15–16 of this manual. Then the exponential regression function can be graphed along with the scatterplot. It can also be evaluated using one of the methods on pages 12–13. The graph shown is in the window [0, 10, 0, 15].

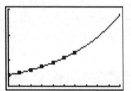

It is also possible to find an exponential regression without using the STAT WIZARD. If the STAT WIZARD is off, press [STAT] [▸] [0]. This pastes ExpReg on the home screen. Then press [2nd] [L1] [,] [2nd] [L2] [,] [VARS] [▸] [1] [1] [ENTER] to find the exponential regression using the x-values in L_1 and the y-values in L_2, and then store the regression equation in Y_1.

Section 5.6, Exercise 30 (a) *Forgetting*. In an economics class, students were given a final exam at the end of the course. Then they were retested with an equivalent test at subsequent time intervals. Their scores after time x, in months, are given in the following table.

Time, x (in months)	Score, y
1	84.9%
2	84.6
3	84.4
4	84.2
5	84.1
6	83.9

Use a graphing calculator to fit a logarithmic function $y = a + b \ln x$ to the data.

After entering the data in lists as described on pages 14–15 of this manual, select LnReg from the STAT CALC menu by pressing [STAT] [▶] [9]. As described earlier, the STAT WIZARD window will open, prompting for the required and optional arguments for the LnReg function. The values of a and b for the logarithmic function $y = a + b \ln x$ are displayed. This function can be evaluated using one of the methods described on pages 12 and 13 of this manual.

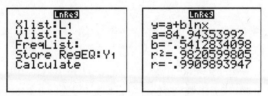

It is also possible to find a logarithmic regression equation without using the STAT WIZARD. If the STAT WIZARD is off, press [STAT] [▶] [9]. This pastes LnReg on the home screen. Then press [2nd] [L1] [,] [2nd] [L2] [,] [VARS] [▶] [1] [1] [ENTER] to find the logarithmic regression equation using the x-values in L_1 and the y-values in L_2, and then store the regression equation in Y_1.

```
LnReg L1,L2,Y1
```

LOGISTIC REGRESSION

A logistic function can be fit to data using the TI-84 Plus.

Section 5.6, Exercise 34 (a) *Effect of Advertising*. A company introduced a new software product on a trial run in a city. They advertised the product on television and found the following data regarding the percent P of people who bought the product after x ads were run.

Number of Ads, x	Percentage Who Bought, P
0	0.2%
10	0.7
20	2.7
30	9.2
40	27.0
50	57.6
60	83.3
70	94.8
80	98.5
90	99.6

Use a graphing calculator to fit a logistic function $P(x) = \dfrac{a}{1 + be^{-kx}}$ to the data.

After entering the data in lists as described on pages 14–15 of this manual, select Logistic from the STAT CALC menu by pressing [STAT] [▸] [ALPHA] [B]. As described earlier, the STAT WIZARD window will open, prompting for the required and optional arguments for the Logistic function. The values of a, b, and c for the logistic function $y = \dfrac{c}{1 + ae^{-bx}}$ are displayed. This function can be evaluated using one of the methods described on pages 12–13 of this manual.

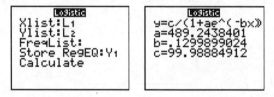

As with the exponential and logarithmic regressions, it is possible to find a logistic regression equation without using the STAT WIZARD. If the STAT WIZARD is off, press [STAT] [▶] [ALPHA] [B]. This pastes Logistic on the home screen. Then press [L1] [,] [L2] [,] [VARS] [▶] [1] [1] [ENTER] to find the logistic regression using the x-values in L_1 and the y-values in L_2, and then store the regression equation in Y_1.

A note about the variables Don't confuse the variables a and b in the formula given in the text with the variables a and b computed on your calculator. The calculator uses the formula $y = \dfrac{c}{1 + ae^{-bx}}$ as the logistic function. This is exactly the same as the function

$P(x) = \dfrac{a}{1 + be^{-kx}}$ given in the text; it just uses different variables.

Chapter 6 Systems of Equations and Matrices

MATRICES AND ROW-EQUIVALENT OPERATIONS

Matrices with up to 99 rows or columns can be entered on a graphing calculator. The TI-84 can store up to 10 matrices. A matrix can be defined directly in an expression using the MTRX shortcut menu. Only real numbers can be stored in TI-84 Plus matrices. Fractions may be entered in fraction form, but are stored as decimals and can be used in matrices. Using MathPrint, it is possible display fractions. Row-equivalent operations can be performed on matrices on the calculator.

Section 6.3, Example 1 Solve the following system.

$$
\begin{aligned}
2x - y + 4z &= -3, \\
x - 2y - 10z &= -6, \\
3x \phantom{{}- 2y} + 4z &= 7.
\end{aligned}
$$

First we will enter the augmented matrix

$$
\left[\begin{array}{ccc|c}
2 & -1 & 4 & -3 \\
1 & -2 & -10 & -6 \\
3 & 0 & 4 & 7
\end{array} \right]
$$

in the graphing calculator. Begin by pressing ⟨2nd⟩ ⟨MATRIX⟩ ⟨▶⟩ ⟨▶⟩ to display the MATRIX EDIT menu. (MATRIX is the second operation associated with the ⟨x^{-1}⟩ key.) Then select the matrix to be defined. We will select matrix [**A**] by pressing 1 or ⟨ENTER⟩. Now the MATRIX EDIT screen appears. The dimensions of the matrix are displayed on the top line of this screen, with the cursor on the row dimension. Enter the dimensions of the augmented matrix, 3 × 4, by pressing ⟨3⟩ ⟨ENTER⟩ ⟨4⟩ ⟨ENTER⟩. Now the cursor moves to the element in the first row and first column of the matrix. Enter the elements of the first row by pressing ⟨2⟩ ⟨ENTER⟩ ⟨(-)⟩ ⟨1⟩ ⟨ENTER⟩ ⟨4⟩ ⟨ENTER⟩ ⟨(-)⟩ ⟨3⟩ ⟨ENTER⟩. The cursor moves to the element in the second row and first column of the matrix. Enter the elements of the second and third rows of the augmented matrix by typing each in turn followed by ⟨ENTER⟩ as above. Note that the screen displays only three columns of the matrix. The arrow keys can be used to move the cursor to any element at any time.

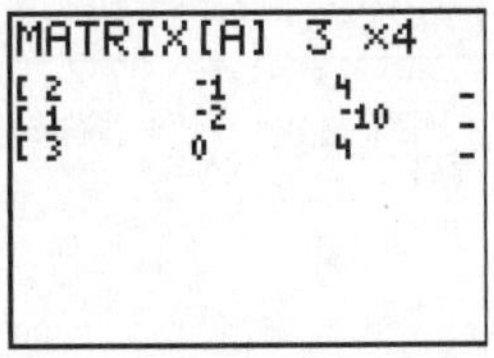

```
MATRIX[A] 3 ×4
[ 2      -1      4       :
[ 1      -2      -10     :
[ 3      0       4       :
```

In MathPrint mode, we can also use the MTRX shortcut menu to enter a matrix. From the home screen , press [ALPHA] [F3] to display the menu.

```
ROW:1 2 3 4 5 6
COL:1 2 3 4 5 6
        OK
FRAC FUNC MTRX YVAR
```

The default matrix size is two rows by two columns. However, if the shortcut has been used before, the calculator won't necessarily show the default setting. Press [▶] or [◀] as many times as necessary followed by [ENTER] to change the number of rows to 3. Then press [▼] followed by [▶] or [◀] as many times as necessary followed by [ENTER] to change the number of columns to four.

```
ROW:1 2 3 4 5 6
COL:1 2 3 4 5 6
        OK
FRAC FUNC MTRX YVAR
```

Press [▼] [ENTER] to display a matrix on the home screen.

```
[0 0 0 0]
[0 0 0 0]
[0 0 0 0]
```

Now fill in each element of the augmented matrix as follows. Press [2] [▶] [(-)] [1] [▶] [4] [▶] [(-)] [3] [▶] [1] [▶] [(-)] [2] [▶] and so on until all of the elements have been entered. Do **not** press [ENTER] after the last entry; press [▶] instead.

```
[2  -1   4  -3]       [2  -1   4  -3]
[1  -2   0   0]       [1  -2 -10  -6]
[0   0   0   0]       [3   0   4   7]
```

Save the matrix as matrix [**A**] by pressing [STO▶] [2nd] [MATRIX] [1] [ENTER].

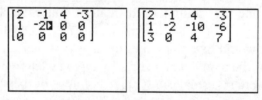

Row-equivalent operations are performed by making selections from the MATRIX MATH menu. To view this menu press [2nd] [QUIT] to leave the MATRIX EDIT screen. (If you've used the shortcut to enter the matrix, then this is not necessary.) Then press [2nd] [MATRIX] [▶]. The MATRIX MATH menu contains 16 operations. The four row-equivalent operations are C: rowSwap(, D: row+(, E: *row(, and F: *row+(. These operations interchange two rows of a matrix, add two rows, multiply a row by a number, and multiply a row by a number and add it to a second row, respectively. Access these operations by pressing [▼] or [▲] until you reach them. Alternatively, you can press [ALPHA] [C] to select rowSwap, [ALPHA] [D] to select row+(, [ALPHA] [E] to select *row(, or [ALPHA] [F] to select *row(+.

We will use the calculator to perform the row-equivalent operations that were done algebraically in the text. First, to interchange row 1 and row 2 of matrix [**A**], with the MATRIX MATH menu displayed, press [ALPHA] [C] to select rowSwap. Then press [2nd] [MATRIX] [1] or [2nd] [MATRIX] [ENTER] to select [**A**]. Follow this with a comma and the rows to be interchanged, [,] [1] [,] [2] [)] [ENTER].

```
rowSwap([A],1,2)
[1  -2  -10  -6]
[2  -1   4   -3]
[3   0   4    7]
```

The graphing calculator will not store the matrix produced using a row-equivalent operation, so when several operations are to be performed in succession it is helpful to store the result of each operation as it is produced. For example, to store the matrix resulting from interchanging the first and second rows of [**A**] as matrix [**B**], press [STO▶] [2nd] [MATRIX] [2] [ENTER] immediately after interchanging the rows. This can also be done before [ENTER] is pressed at the end of the rowSwap.

Next we multiply the first row of [**B**] by −2, add it to the second row, and store the result as [**B**] again by pressing [2nd] [MATRIX] [▶] [ALPHA] [F] [(-)] [2] [,] [2nd] [MATRIX] [2] [,] [1] [,] [2] [)] [STO▶] [2nd] [MATRIX] [2] [ENTER]. These keystrokes select F: *row+(from the MATRIX MATH menu; then they specify that the value of the multiplier is −2, the matrix being operated on is [**B**], and that a multiple of row 1 is being added to row 2; finally they store the result as [**B**].

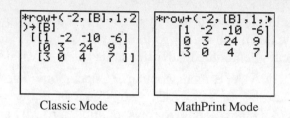

Classic Mode   MathPrint Mode

To multiply row 1 by −3 and then add it to row 3, use F:*row+(again. We will store the result as [**B**]. Press 2nd [MATRIX] ▶ ALPHA [F] (-) 3 , 2nd [MATRIX] 2 , 1 , 3) STO▶ 2nd [MATRIX] 2 ENTER.

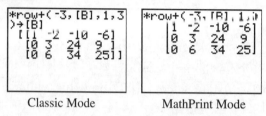

Classic Mode   MathPrint Mode

Now multiply the second row by $\frac{1}{3}$ using E:*row(, and store the result as [**B**] again.

Press 2nd [MATRIX] ▶ ALPHA [E] 1 ÷ 3 , 2nd [MATRIX] 2 , 2) STO▶ 2nd [MATRIX] 2 ENTER. These keystrokes select E: *row(from the MATRIX MATH menu; then they specify that the value of the multiplier is $\frac{1}{3}$, the matrix being operated on is [**B**], and row 2 is being multiplied; finally they store the result as [**B**]. The keystrokes 1 ÷ 3 could be replaced with 3 x^{-1}.

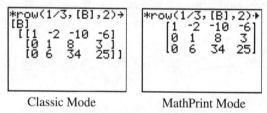

Classic Mode   MathPrint Mode

Next multiply the second row by −6 and add it to the third row using F: *row+(. Press 2nd [MATRIX] ▶ ALPHA [F] (-) 6 , 2nd [MATRIX] 2 , 2 , 3) STO▶ 2nd [MATRIX] 2 ENTER.

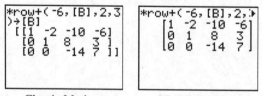

Classic Mode   MathPrint Mode

Finally, multiply the third row by $-\frac{1}{14}$ using E:*row(by pressing [2nd] [MATRIX] [▶] [ALPHA] [E] [(-)] [1] [÷] [1] [4] [,] [2nd] [MATRIX] [2] [,] [3] [)] [ENTER]. The keystrokes [(-)] [1] [÷] [1] [4] could be replaced with [(-)] [1] [4] [x⁻¹].

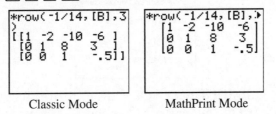

Classic Mode   MathPrint Mode

Write the system of equations that corresponds to the final matrix. Then use back-substitution to solve for x, y, and z as illustrated in the text.

Instead of stopping with row-echelon form as we did above, we can continue to apply row-equivalent operations until the matrix is in reduced row-echelon form as in **Example 3** in **Section 6.3** of the text. Reduced row-echelon form of a matrix can be found directly by using the rref(operation from the MATRIX MATH menu. For example, to find reduced row-echelon form for matrix **A** in Example 1 above, after entering [A] (and leaving the MATRIX EDIT screen if you haven't used the shortcut), press [2nd] [MATRIX] [▶] [ALPHA] [B] [2nd] [MATRIX] [1] [)] [ENTER]. We can read the solution of the system of equations, $(3, 7, -0.5)$ directly from the resulting matrix.

```
rref([A])
        [1 0 0  3 ]
        [0 1 0  7 ]
        [0 0 1 -.5]
```

USING THE PolySmlt2 APPLICATION

The application PolySmlt2 from the APPS menu can be used to solve a system of equations. If your calculator does not have this application pre-loaded, you can download the Polynomial Root Finder and Simultaneous Equation Solver version 2 from the Texas Instruments website, http://education.ti.com, or have it transmitted to your calculator from a calculator that contains the App.

We will do **Example 1** in **Section 6.3** again using the PolySmlt2 App. Solve the following system.

$$
\begin{aligned}
2x &- y + 4z = -3, \\
x &- 2y - 10z = -6, \\
3x & \quad\;\; + 4z = 7.
\end{aligned}
$$

To access PolySmlt2, press the [APPS] key, scroll down to PolySmlt2 (pressing [ALPHA] [P] will move the cursor to the first "p" entry), and press [ENTER]. Next press any key to display the Main Menu.

```
        MAIN MENU
1: POLY ROOT FINDER
2: SIMULT EQN SOLVER
3: ABOUT
4: POLY HELP
5: SIMULT HELP
6: QUIT POLYSMLT
```

Press [2] to select item 2, SIMULT EQN SOLVER. Press the right or left arrow key followed by [ENTER] to select the number of equations, and then press [▾] followed by the right or left arrow keys followed by [ENTER] again to select the number of unknowns, or variables. Access the next screen, a template for an augmented matrix, by pressing [GRAPH] key, which is located directly under NEXT on the solver screen. Enter the augmented matrix for this system of equations as described on page 45 of this manual. Finally, press the key directly below the "SOLVE" option at the bottom of the screen, the [GRAPH] key. The solution (3, 7, −0.5) is displayed.

```
  SIMULT EQN SOLVER MODE          SYSTEM MATRIX (3×4)              SOLUTION
EQUATIONS2 3 4 5 6 7 8 9 10     [2   -1    4   -3 ]          x1 =3
UNKNOWNS  2 3 4 5 6 7 8 9 10    [1   -2  -10   -6 ]          x2 =7
DEC       FRAC                  [3    0    4    7 ]          x3 = -1/2
NORMAL    SCI ENG
FLOAT     0 1 2 3 4 5 6 7 8 9   (3,4)=7
RADIAN    DEGREE
(MAIN)          (HELP)(NEXT)    (MAIN)(MODE)(CLR)(LOAD)(SOLVE)   (MAIN)(MODE)(SYSM)(STO)(F◄►D)
```

Press [2nd] [QUIT] or [Y=], which is located under MAIN on the solver screen followed by [6] to exit the app.

MATRIX OPERATIONS

We can use a graphing calculator to add and subtract matrices, to multiply a matrix by a scalar, and to multiply matrices.

Section 6.4, Example 1 (a) Find **A** + **B** for

$$\mathbf{A} = \begin{bmatrix} -5 & 0 \\ 4 & \frac{1}{2} \end{bmatrix}, \quad \mathbf{B} = \begin{bmatrix} 6 & -3 \\ 2 & 3 \end{bmatrix}.$$

Enter **A** and **B** on the MATRIX EDIT screen as [A] and [B] as described earlier in this chapter of the Graphing Calculator Manual. Press [2nd] [QUIT] to leave this screen. Then press [2nd] [MATRIX] [1] [+] [2nd] [MATRIX] [2] [ENTER] to display the sum.

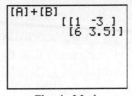

Classic Mode

Alternatively, we can use the MTRX shortcut to enter and then add the matrices without storing them. Be sure that your calculator is in MathPrint mode. Press ALPHA [F3] to display the menu and fill in the elements of matrix **A** as described on page 46 of this manual. Then press + ALPHA [F3]. Now enter the elements in matrix **B** as described earlier. Finally, press ENTER. The sum of the matrices is displayed.

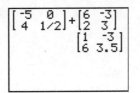

Section 6.4, Example 2 Find **C** − **D** for each of the following.

a) $C = \begin{bmatrix} 1 & 2 \\ -2 & 0 \\ -3 & -1 \end{bmatrix}$, $D = \begin{bmatrix} 1 & -1 \\ 1 & 3 \\ 2 & 3 \end{bmatrix}$ b) $C = \begin{bmatrix} 5 & -6 \\ -3 & 4 \end{bmatrix}$, $D = \begin{bmatrix} -4 \\ 1 \end{bmatrix}$

Enter **C** and **D** on the MATRIX EDIT screen as [**C**] and [**D**]. Press 2nd [QUIT] to leave this screen. Then press 2nd [MATRIX] 3 − 2nd [MATRIX] 4 ENTER to display the difference.

```
[C]-[D]
         [ 0   3 ]
         [-3  -3 ]
         [-5  -4 ]
```

As before, we can use the MTRX shortcut to enter and then subtract the matrices without storing them.

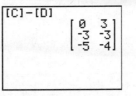

b) Enter **C** and **D** on the MATRIX EDIT screen as [**C**] and [**D**]. Press 2nd [QUIT] to leave this screen. Then press 2nd [MATRIX] 3 − 2nd [MATRIX] 4 [ENTER]. The calculator returns the message ERR:DIM MISMATCH, indicating that this subtraction is not possible. This is the case because the matrices have different orders. We also obtain an error message if we use the MTRX shortcut method.

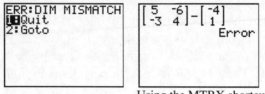

Using the MTRX shortcut

Section 6.4, Example 4 Find 3**A** and (−1)**A**, for

$$\mathbf{A} = \begin{bmatrix} -3 & 0 \\ 4 & 5 \end{bmatrix}.$$

Enter **A** on the MATRIX EDIT screen as [**A**]. Press 2nd [QUIT] to leave this screen. Then to find 3**A** press 3 2nd [MATRIX] 1 (or [ENTER]) [ENTER] and to find (−1)**A** press (−) 1 2nd [MATRIX] 1 (or [ENTER]) [ENTER]. Note that (−1)**A** is the opposite, or additive inverse, of **A** and can also be found by pressing (−) 2nd [MATRIX] 1 (or [ENTER]) [ENTER].

```
3[A]
              [-9   0]
              [12  15]
-1[A]
              [ 3   0]
              [-4  -5]
```

It is possible to use the MTRX shortcut to perform these operations. However, it is not quite as convenient because the matrix is not stored, and must be reentered after calculating 3**A**.

```
 [-3  0]
3[ 4  5]
              [-9   0]
              [12  15]
```
```
  [-3  0]
-1[ 4  5]
              [ 3   0]
              [-4  -5]
```

Section 6.4, Example 6 (a), (d) For

$$A = \begin{bmatrix} 3 & 1 & -1 \\ 2 & 0 & 3 \end{bmatrix}, \ B = \begin{bmatrix} 1 & 6 \\ 3 & -5 \\ -2 & 4 \end{bmatrix}, \text{ and } C = \begin{bmatrix} 4 & -6 \\ 1 & 2 \end{bmatrix}$$

find each of the following.

a) AB   **d) AC**

First enter **A**, **B**, and **C** as [A], [B], and [C] on the MATRIX EDIT screen. Press [2nd] [QUIT] to leave this screen.

a) To find **AB**, press [2nd] [MATRIX] [1] [2nd] [MATRIX] [2] [ENTER].

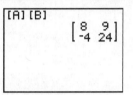

It is possible to use the MTRX shortcut to perform these operations. However, matrix **A** must be entered again in order to find **AC** in part (d) since it was not stored.

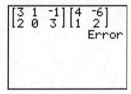

d) To find **AC**, press [2nd] [MATRIX] [1] [2nd] [MATRIX] [3] [ENTER]. The calculator returns the message ERR:DIM MISMATCH, indicating that this multiplication is not possible. This is the case because the number of columns in **A** is not the same as the number of rows in **C**. Thus, the matrices cannot be multiplied in this order. The screen below shows the error on the home screen when the matrices are entered using the MTRX shortcut.

Using the MTRX shortcut

FINDING THE INVERSE OF A MATRIX

The inverse of a matrix can be found quickly on a graphing calculator.

Section 6.5, Example 3 Find A^{-1}, where

$$A = \begin{bmatrix} -2 & 3 \\ -3 & 4 \end{bmatrix}.$$

Enter **A** as [A] on the MATRIX EDIT screen. Then press 2nd [QUIT] to leave this screen. Now press 2nd [MATRIX] 1 x^{-1} ENTER to display A^{-1}. We can also find the inverse by entering the matrix using the MTRX shortcut followed by x^{-1} ENTER.

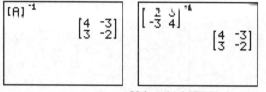

Using the MTRX shortcut

Section 6.5, Exercise 7 Find A^{-1}, where

$$A = \begin{bmatrix} 6 & 9 \\ 4 & 6 \end{bmatrix}.$$

Enter **A** as [A] on the MATRIX EDIT screen. Then press 2nd [QUIT] to leave this screen. Now press 2nd [MATRIX] 1 x^{-1} ENTER. The calculator returns the message ERR:SINGULAR MAT, indicating that A^{-1} does not exist.

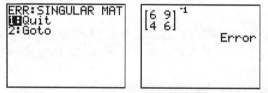

Using the MTRX shortcut

MATRIX SOLUTIONS OF SYSTEMS OF EQUATIONS

We can write a system of n linear equations in n variables as a matrix equation $\mathbf{AX} = \mathbf{B}$. If $\mathbf{A}$ has an inverse, the solution of the system of equations is given by $\mathbf{X} = \mathbf{A}^{-1}\mathbf{B}$.

Section 6.5, Example 5 Use an inverse matrix to solve the following system of equations.

$$-2x + 3y = 4$$
$$-3x + 4y = 5.$$

Enter $\mathbf{A} = \begin{bmatrix} -2 & 3 \\ -3 & 4 \end{bmatrix}$ and $\mathbf{B} = \begin{bmatrix} 4 \\ 5 \end{bmatrix}$ on the MATRIX EDIT screen as [A] and [B]. Press

[2nd] [QUIT] to leave this screen. Then press [2nd] [MATRIX] [1] (or [ENTER]) [x⁻¹] [2nd] [MATRIX] [2]

[ENTER]. The result is the 2×1 matrix $\begin{bmatrix} 1 \\ 2 \end{bmatrix}$, so the solution is (1, 2). As with the previous

examples, the MTRX shortcut can be used to find the solution.

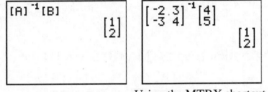

Using the MTRX shortcut

DETERMINANTS AND CRAMER'S RULE

We can evaluate determinants on a graphing calculator and use Cramer's rule to solve systems of equations.

Section 6.6, Example 5 Use a graphing calculator to evaluate $|\mathbf{A}|$.

$$\mathbf{A} = \begin{bmatrix} 1 & 6 & -1 \\ -3 & -5 & 3 \\ 0 & 4 & 2 \end{bmatrix}.$$

Enter $\mathbf{A}$ on the MATRIX EDIT screen and then press [2nd] [QUIT] to leave this screen. We will select the "det("operation from the MATRIX MATH menu. Press [2nd] [MATRIX] [▶] [1] (or [ENTER]) [2nd] [MATRIX] [1] (or [ENTER]) [)] [ENTER]. We see that $|\mathbf{A}|=26$.

Alternatively, we can use the MTRX shortcut. Press ⎨2nd⎬ [MATRIX] ⎨▶⎬ ⎨1⎬ ⎨ALPHA⎬ [F3] followed by the matrix elements (as described on page 46 of this manual). Then press ⎨)⎬ ⎨ENTER⎬.

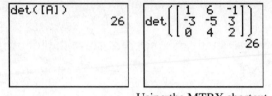

Using the MTRX shortcut

Section 6.6, Example 6 Solve using Cramer's rule.

$$2x + 3y = 7,$$
$$5x - 2y = -3$$

First we enter the matrices corresponding to D, D_x, and D_y as **A**, **B**, and **C**, respectively. We have

$$\mathbf{A} = \begin{bmatrix} 2 & 5 \\ 5 & -2 \end{bmatrix}, \quad \mathbf{B} = \begin{bmatrix} 7 & 5 \\ -3 & -2 \end{bmatrix}, \text{ and } \mathbf{C} = \begin{bmatrix} 2 & 7 \\ 5 & -3 \end{bmatrix}.$$

Then we use the "det(" operation from the MATRIX MATH menu. We have

$$x = \frac{\det(\mathbf{B})}{\det(\mathbf{A})} \text{ and } y = \frac{\det(\mathbf{C})}{\det(\mathbf{A})}.$$

To find x expressed in fraction notation, press ⎨2nd⎬ [MATRIX] ⎨▶⎬ ⎨1⎬ (or ⎨ENTER⎬) ⎨2nd⎬ [MATRIX] ⎨2⎬ ⎨)⎬ ⎨÷⎬ ⎨2nd⎬ [MATRIX] ⎨▶⎬ ⎨1⎬ (or ⎨ENTER⎬) ⎨2nd⎬ [MATRIX] ⎨1⎬ (or ⎨ENTER⎬) ⎨)⎬ ⎨MATH⎬ ⎨1⎬ (or ⎨ENTER⎬) ⎨ENTER⎬. The result is $-\frac{1}{29}$. To find y expressed in fraction notation, press ⎨2nd⎬ [MATRIX] ⎨▶⎬ ⎨1⎬ (or ⎨ENTER⎬) ⎨2nd⎬ [MATRIX] ⎨3⎬ ⎨)⎬ ⎨÷⎬ ⎨2nd⎬ [MATRIX] ⎨▶⎬ ⎨1⎬ (or ⎨ENTER⎬) ⎨2nd⎬ [MATRIX] ⎨1⎬ (or ⎨ENTER⎬) ⎨)⎬ ⎨MATH⎬ ⎨1⎬ (or ⎨ENTER⎬) ⎨ENTER⎬. To find y, we can also recall the entry for x and edit it, replacing matrix **B** with matrix **C**. In either case the result is $\frac{41}{29}$. The solution of the system of equations is $\left(-\frac{1}{29}, \frac{41}{29}\right)$.

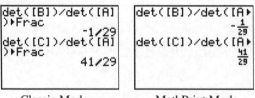

Classic Mode   MathPrint Mode

GRAPHS OF INEQUALITIES

We can graph linear inequalities on a graphing calculator, shading the region of the solution set. The calculator should be set in Func mode.

Section 6.7, Example 1 Graph $y < x + 3$.

Clear any existing drawings by pressing $\boxed{\text{2nd}}$ $\boxed{\text{DRAW}}$ $\boxed{1}$ $\boxed{\text{ENTER}}$. Now graph the related equation $y = x + 3$. We will use the standard window $[-10, 10, -10, 10]$. Since the inequality symbol is $<$, we know that the line $y = x + 3$ is not part of the solution set. In a hand-drawn graph we would use a dashed line to indicate this. We might try selecting Dot mode or use the Dot GraphStyle, but even if Dot mode is used, the line appears to be solid. After determining that the solution set of the inequality consists of all points below the line, we use the calculator's SHADE operation to shade this region. SHADE is item 7 on the DRAW DRAW menu. To access it press $\boxed{\text{2nd}}$ $\boxed{\text{DRAW}}$ $\boxed{7}$.

Now enter a lower function and an upper function. The region between them will be shaded. We want to shade the area between the bottom of the window, $y = -10$, and the line $y = x + 3$, so we enter $\boxed{\text{(-)}}$ $\boxed{1}$ $\boxed{0}$ $\boxed{,}$ $\boxed{\text{X,T,Θ,}n}$ $\boxed{+}$ $\boxed{3}$ $\boxed{)}$ $\boxed{\text{ENTER}}$. We can also enter $x + 3$ as Y_1. The result is shown below. Keep in mind that the line $y = x + 3$ is not included in the solution set.

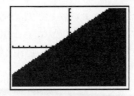

The "shade below" GraphStyle can also be used to shade this region. After entering the related equation, cycle through the Graph Style options on the "Y =" screen by positioning the cursor over the GraphStyle icon to the left of the equation and pressing $\boxed{\text{ENTER}}$ repeatedly until the "shade below" icon appears. If the "line" icon was previously selected, $\boxed{\text{ENTER}}$ must be pressed three times to select the "shade below" style. Then press $\boxed{\text{GRAPH}}$ to display the graph of the inequality. As mentioned above, keep in mind the fact that the line $y = x + 3$ is not included in the solution set.

Plot1 Plot2 Plot3
\Y₁⬛X+3
\Y₂=
\Y₃=
\Y₄=
\Y₅=
\Y₆=
\Y₇=

Some calculators have a pre-loaded application, Inequalz, on the APPS menu that can be used to graph inequalities. If your calculator does not have this App, you can download the Inequality Graphing App version 1.04 from the Texas Instruments website, http://education.ti.com, or have it transmitted to your calculator from a calculator that contains it. To use Inequalz to graph $y < x + 3$, first press the $\boxed{\text{APPS}}$ key, press $\boxed{\text{ALPHA}}$ [i] to jump to the i's, then scroll down to Inequalz, and press $\boxed{\text{ENTER}}$. Then press any key to display the Y= screen. Enter $Y_1 < x + 3$ by first pressing $\boxed{\text{ALPHA}}$ [F2] to select the inequality symbol < from the menu at the bottom of the screen. (F2 is the Alpha operation associated with the $\boxed{\text{WINDOW}}$ key. This key is used because it lies directly below the < symbol on the screen.)

Now press $\boxed{\blacktriangleright}$ to position the cursor to the right of Y_1 < and continue by pressing $\boxed{\text{X,T,}\Theta,n}$ $\boxed{1}$ $\boxed{3}$. Then select a window and press $\boxed{\text{GRAPH}}$ to see the graph of the inequality. Here we pressed $\boxed{\text{ZOOM}}$ $\boxed{6}$ to select the standard window. Note that the App has the capability to graph the related equation, $y = x + 3$, as a dashed line.

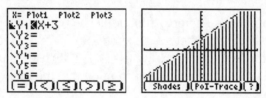

To quit the Inequalz App, select Inequalz from the APPS menu again and then press $\boxed{2}$ to select Quit from the INEQUAL RUNNING menu. Additional information about this App can be found at http://education.ti.com/downloads/guidebooks/apps/83inequality_graphing/ineq-eng.pdf.

We can also use the Inequalz App to graph an inequality with a related equation of the form $x = a$.

Section 6.7, Example 3 Graph $x > -3$ on a plane.

Select the Inequalz App and display the Y= screen as described above. Clear any entries currently on the screen. Then press $\boxed{\blacktriangle}$ to position the cursor on X = at the top of the screen and press $\boxed{\text{ENTER}}$. Then enter $X_1 > -3$ by first pressing $\boxed{\text{ALPHA}}$ [F4] to select the inequality symbol >. (F4 is the Alpha operation associated with the $\boxed{\text{TRACE}}$ key. It is the key directly below the > symbol on the screen.) Next press $\boxed{\blacktriangleright}$ to position the cursor to the right of $X_1 >$ and continue to enter the inequality by pressing $\boxed{(-)}$ $\boxed{3}$. Then select a window and press $\boxed{\text{GRAPH}}$ to see the graph of the inequality. The standard window is shown on the next page.

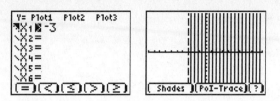

Recall that, to quit the Inequalz App, select Inequalz from the APPS menu again and then press $\boxed{2}$ to select Quit.

We can use the SHADE operation to graph a system of inequalities when the solution set lies between the graphs of two functions.

Section 6.7, Exercise 43 Graph

$$y \leq x,$$
$$y \geq 3 - x.$$

First graph the related equations $y_1 = x$ and $y_2 = 3 - x$ and determine that the solution set consists of all the points on or above the graph of $y_2 = 3 - x$ and on or below the graph of $y_1 = x$. Clear any existing drawings by pressing $\boxed{\text{2nd}}$ $\boxed{\text{DRAW}}$ $\boxed{1}$ $\boxed{\text{ENTER}}$. Then shade this region by pressing $\boxed{\text{2nd}}$ $\boxed{\text{DRAW}}$ $\boxed{7}$ $\boxed{3}$ $\boxed{-}$ $\boxed{\text{X,T,}\Theta,n}$ $\boxed{,}$ $\boxed{\text{X,T,}\Theta,n}$ $\boxed{)}$ $\boxed{\text{ENTER}}$. These keystrokes select the SHADE operation from the DRAW DRAW menu and then enter $y_2 = 3 - x$ as the lower function and $y_1 = x$ as the upper function. We could also enter these functions as Y_2 and Y_1, respectively, in the Shade() instruction by pressing $\boxed{\text{2nd}}$ $\boxed{\text{DRAW}}$ $\boxed{7}$ $\boxed{\text{VARS}}$ $\boxed{\blacktriangleright}$ $\boxed{1}$ (or $\boxed{\text{ENTER}}$) $\boxed{\blacktriangledown}$ $\boxed{\text{ENTER}}$ $\boxed{,}$ $\boxed{\text{VARS}}$ $\boxed{\blacktriangleright}$ $\boxed{1}$ (or $\boxed{\text{ENTER}}$) $\boxed{\text{ENTER}}$ $\boxed{)}$ $\boxed{\text{ENTER}}$.

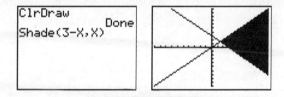

We can also graph systems of inequalities by shading the solution set of each inequality in the system with a different pattern. When the "shade above" or "shade below" GraphStyle options are selected the calculator rotates through four shading patterns. Vertical lines shade the first function, horizontal lines the second, negatively sloping diagonal lines the third, and positively sloping diagonal lines the fourth. These patterns repeat if more than four functions are graphed. The region of overlap is the solution set of the system of inequalities.

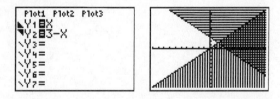

Section 6.7, Example 5 Graph the solution set of the system

$$x + y \leq 4$$
$$x - y \geq 2.$$

First graph the equation $x + y = 4$, entering it in the form $y_1 = -x + 4$. We determine that the solution set of $x + y \leq 4$ consists of all points below the line $x + y = 4$, or $y_1 = -x + 4$, so we select the "shade below" GraphStyle for this function. Next graph $x - y = 2$, entering it in the form $y_1 = x - 2$. The solution set of $x - y \geq 2$ is all points below the line $x - y = 2$, or $y_2 = x - 2$, so we also choose the "shade below" GraphStyle for this function. Now press GRAPH to display the solution sets of each inequality in the system and the region where they overlap. The region of overlap is the solution set of the system of inequalities.

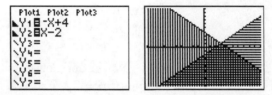

We can also use the Inequalz App to graph this system of inequalities. First we solve each inequality for y, obtaining $y \leq -x + 4$ from $x + y \leq 4$ and $y \leq x - 2$ from $x - y \geq 2$. Press APPS, select the Inequalz App, and enter $Y_1 \leq -x + 4$ and $Y_2 \leq x - 2$ as described on page 58 of this manual. Then select a viewing window and press GRAPH. Here we pressed ZOOM 6 to select the standard window. Initially we see a graph like the one above, with the solution sets of both inequalities shaded. To show only the solution set of the system of inequalities, press ALPHA [F1] to select Shades. (F1 is the Alpha operation associated with the Y= key.) Then press 1 to select Intersection and display the graph shown below.

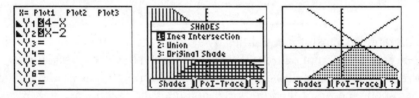

FINDING THE COORDINATES OF VERTICES

If a system of inequalities is graphed using a shading option, the coordinates of the vertices can be found using the Intersect feature. If a system of inequalities is graphed using the Inequalz application from the APPS menu, the PoI-Trace (point of intersection) feature can be used to find the coordinates of the vertices.

Section 6.7, Example 6 Graph the following system of inequalities and find the coordinates of any vertices formed.

$$3x - y \leq 6 \qquad (1)$$
$$y - 3 \leq 0 \qquad (2)$$
$$x + y \geq 0 \qquad (3)$$

We will demonstrate the use of the PoI-Trace feature. First we access the Inequalz App and graph the three inequalities as described on page 58 of this manual. To do this we first solve each inequality for y and graph $Y_1 \geq 3x - 6$, $Y_2 \leq 3$, and $Y_3 \geq -x$. In the graphs shown below we have used the window $[-5, 5, -5, 5]$ and we have also used the Shades feature to shade only the solution set of the system of inequalities. The use of Shades is described previously in Example 5.

To find the coordinates of the vertices, first press [ALPHA] [F3] to select the PoI-Trace option at the bottom of the screen. (F3 is the Alpha option associated with the [ZOOM] key.) We see the coordinates of the vertex (3, 3) displayed at the bottom of the screen. The notation $Y1 \cap Y2$ at the top of the screen tells us that this is the vertex associated with the system of equations from inequalities (1) and (2) above. Press the [▼] key twice to display the coordinates of the vertex associated with inequalities (1) and (3), (1.5, −1.5), and press [◄] to find the vertex associated with (2) and (3), (−3, 3).

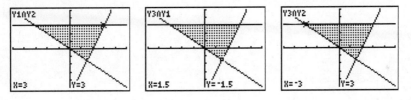

Chapter 7 Conic Sections

Many conic sections are represented by equations that are not functions. Consequently, these equations must be entered on the equation-editor screen as two equations, each of which is a function. Alternatively, some calculators have a pre-loaded application that allows conic sections to be graphed in standard form. Unless otherwise noted, we use Classic mode throughout this chapter so that text wraps rather than scrolls off the screen.

GRAPHING PARABOLAS

To graph a parabola of the form $y^2 = 4px$ or $(y-k)^2 = 4p(x-h)$, we must first solve the equation for y.

Section 7.1, Example 4 Graph the parabola $y^2 - 2y - 8x - 31 = 0$.

One way to produce the graph is to solve the equation for y. In the text, we completed the square to obtain $(y-1)^2 = 8x + 32$. Now solve for y.

$$(y-1)^2 = 8x + 32$$
$$y - 1 = \pm\sqrt{8x + 32} \qquad \text{Taking the square root on both sides.}$$
$$y = 1 \pm \sqrt{8x + 32} \qquad \text{Adding 1 on both sides.}$$

Enter $Y_1 = 1 + \sqrt{8x + 32}$ and $Y_2 = 1 - \sqrt{8x + 32}$, select a window, and press $\boxed{\text{GRAPH}}$ to see the graph. Here we use the square window $[-12, 12, -8, 8]$. The first equation produces the top half of the parabola and the second equation produces the lower half. Note that the upper and lower halves do not connect because of the way the TI-84 plots a graph.

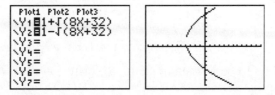

We can also enter $Y_1 = \sqrt{8x + 32}$ and then enter $Y_2 = 1 + Y_1$ and $Y_3 = 1 - Y_1$. To enter $Y_2 = 1 + Y_1$ position the cursor beside "$Y_2 =$" and press $\boxed{1}$ $\boxed{+}$ $\boxed{\text{VARS}}$ $\boxed{\blacktriangleright}$ $\boxed{1}$ $\boxed{1}$. Enter $Y_3 = 1 - Y_1$ in a similar manner. Finally, deselect Y_1 by moving the cursor to the equals sign following Y_1 and pressing $\boxed{\text{ENTER}}$. Note that the equals sign in this equation is no longer highlighted. This indicates that it has been deselected and thus its graph will not appear with the graphs of the equations that remain selected.

The top half of the graph is produced by Y_2 and the lower half by Y_3. The expression for Y_1 was entered to avoid entering the radical expression more than once. By deselecting Y_1, we prevent its graph from appearing on the screen with the graph of the parabola.

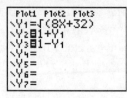

It is also possible to solve the equation for y using the quadratic formula. In the equation $y^2 - 2y - 8x - 31 = 0$, $a = 1$, $b = -2$, and $c = -8x - 31$.

$$y = \frac{-(-2) \pm \sqrt{(-2)^2 - 4(1)(-8x - 31)}}{2(1)} = \frac{2 \pm \sqrt{4 + 32x + 124}}{2} = \frac{2 \pm \sqrt{32x + 128}}{2}$$

As before, enter $Y_1 = \dfrac{2 + \sqrt{32x + 128}}{2}$ and $Y_2 = \dfrac{2 - \sqrt{32x + 128}}{2}$, select a window, and press GRAPH to see the graph. Alternatively, enter $Y_1 = \sqrt{32x + 128}$, and then enter $Y_2 = \dfrac{2 + Y_1}{2}$ and $Y_3 = \dfrac{2 - Y_1}{2}$. To enter $Y_2 = \dfrac{2 + Y_1}{2}$ position the cursor beside "$Y_2 =$" and press (2 + VARS ▶ 1 1) ÷ 2 . Enter $Y_3 = 1 - Y_1$ in a similar manner. Deselect Y_1 as described earlier, then press GRAPH to see the graph.

The Conics application from the APPS menu can also be used to graph a parabola. This App requires the equation to be written in standard form, $(y - k)^2 = 4p(x - h)$ or $(x - h)^2 = 4p(y - k)$. In the text we found standard form for the equation of the parabola $y^2 - 2y - 8x - 31 = 0$ to be $(y - 1)^2 = 4(2)\big[x - (-4)\big]$.

To use the Conics App to graph this parabola, first press APPS, scroll down to Conics, and then press ENTER to see the Conics menu. Next press 4 to select Parabola. Press 1 or ENTER to select an equation in the form $(Y - K)^2 = 4P(X - H)$. We have H = −4, K = 1, and P = 2. Enter these values on the parabola screen by pressing (-) 4 ENTER 1

$\boxed{\text{ENTER}}$ $\boxed{2}$ $\boxed{\text{ENTER}}$. Finally press $\boxed{\text{GRAPH}}$ to graph the parabola in a window selected by the calculator.

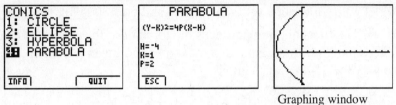

Graphing window
automatically selected

If you wish to select a window manually, press $\boxed{\text{MODE}}$ and select MAN on the WINDOW SETTINGS line. Then press $\boxed{\text{Y=}}$, which corresponds to the ESC (Escape) command, to leave the CONIC SETTINGS screen. Finally, press $\boxed{\text{WINDOW}}$ and enter the desired dimensions.

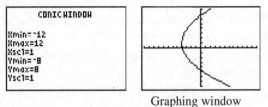

Graphing window
manually selected

Quit the Conics App by pressing $\boxed{\text{2nd}}$ [QUIT].

GRAPHING CIRCLES

The equation of a circle must be solved for y before it can be entered on the equation-editor screen. The Conics App can also be used to graph a circle, if available. We can also use the Circle feature from the DRAW menu, as discussed on pages 11–12 of this manual.

Section 7.2, Example 1 Graph the circle $x^2 + y^2 - 16x + 14y + 32 = 0$.

In the text we found the standard form for the equation of the circle, $(x-8)^2 + \left[y-(-7)\right]^2 = 9^2$. Now we solve for y.

$$(x-8)^2 + \left[y-(-7)\right]^2 = 9^2$$

$$\left[y-(-7)\right]^2 = 81 - (x-8)^2 \qquad 9^2 = 81; \text{ subtracting } (x-8)^2 \text{ on both sides}$$

$$y + 7 = \pm\sqrt{81 - (x-8)^2} \qquad \text{Taking the square root on both sides}$$

$$y = -7 \pm \sqrt{81 - (x-8)^2} \qquad \text{Subtracting 7 on both sides}$$

One way to produce the graph is to enter $Y_1 = -7 + \sqrt{81 - (x - 8)^2}$ and

$Y_2 = -7 - \sqrt{81 - (x - 8)^2}$, select a square window, and press GRAPH. Here we use
[−12, 24, −20, 4]. The first equation produces the top half of the circle and the second
equation produces the lower half. Note that the upper and lower halves do not connect
because of the way the TI-84 plots a graph.

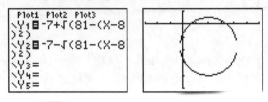

We can also enter $Y_1 = \sqrt{81 - (x - 8)^2}$ and then enter $Y_2 = -7 + Y_1$ and $Y_3 = -7 - Y_1$.
Then deselect Y_1, select a square window, and press GRAPH. We use Y_1 to eliminate the
need to enter the radical expression more than once. Deselecting it prevents the graph of
Y_1 from appearing on the screen with the graph of the circle. The top half of the graph is
produced by Y_2 and the lower half by Y_3.

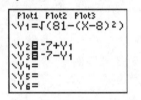

Alternatively, we could have solved the original equation for y using the quadratic
formula, with $a = 1$, $b = 14$, and $c = x^2 - 16x + 32$. This leads to

$$y = \frac{-14 \pm \sqrt{-4x^2 + 64x + 68}}{2}, \text{ or } y = -7 \pm \sqrt{-x^2 + 16x + 17}.$$

To graph the circle $x^2 + y^2 - 16x + 14y + 32 = 0$ using the Conics App, first press APPS,
scroll down to Conics, and then press ENTER to see the Conics menu. Next press 1 or
ENTER to select Circle. We see that we can enter an equation in standard form, $(X - H)^2 +$
$(Y - K)^2 = R^2$, or in the form $AX^2 + AY^2 + BX + CY + D = 0$. Since our equation is
given in the second form, we press 2 to select the second option.

We have A = 1, B = −16, C = 14, and D = 32. Enter these values by pressing 1 ENTER
(-) 1 6 ENTER 1 4 ENTER 3 2 ENTER, then press GRAPH to see the circle. If you
want the window to be automatically selected, be sure that WINDOW SETTINGS on the
CONIC SETTINGS screen (obtained by pressing MODE) is set to AUTO. The window
can be selected manually as described earlier in Section 7.1, Example 4.

Graphing window
automatically selected

Quit the Conics App by pressing [2nd] [QUIT].

GRAPHING ELLIPSES

As with parabolas and circles, the equation of an ellipse must be solved for y before it can be entered on the equation-editor screen. The Conics App can also be used to graph an ellipse.

Section 7.2, Example 2 Find the standard equation of the ellipse with vertices $(-5, 0)$ and $(5, 0)$ and foci $(-3, 0)$ and $(3, 0)$. Then graph the ellipse.

The standard equation for this ellipse, $\dfrac{x^2}{25} + \dfrac{y^2}{16} = 1$, was found on page 588 in the text.

The procedure for graphing an ellipse of the form $\dfrac{x^2}{a^2} + \dfrac{y^2}{b^2} = 1$ or $\dfrac{x^2}{b^2} + \dfrac{y^2}{a^2} = 1$ is similar

to that for graphing a parabola or a circle as described earlier. Solving for y gives

$y = \pm\sqrt{\dfrac{400 - 16x^2}{25}}$. Enter $Y_1 = \sqrt{\dfrac{400 - 16x^2}{25}}$ and $Y_2 = -\sqrt{\dfrac{400 - 16x^2}{25}}$, select a square

window, and press [GRAPH]. Alternatively, we can enter $Y_1 = \sqrt{\dfrac{400 - 16x^2}{25}}$ and $Y_2 = -Y_1$.

Here we used the window $[-9, 9, -6, 6]$.

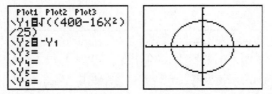

The procedure for using the Conics App that follows on page 69 of this manual can also be used to graph ellipses in this form. Refer also to page 589 in the text.

Now we consider ellipses of the form $\dfrac{(x-h)^2}{a^2}+\dfrac{(y-k)^2}{b^2}=1$ or $\dfrac{(x-h)^2}{b^2}+\dfrac{(y-k)^2}{a^2}=1$.

Section 7.2, Example 4   Graph the ellipse $4x^2 + y^2 + 24x - 2y + 21 = 0$.

Completing the square in the text, we found that the equation can be written as

$$\frac{(x+3)^2}{4}+\frac{(y-1)^2}{16}=1.$$

Solve this equation for y.

$$\frac{(x+3)^2}{4}+\frac{(y-1)^2}{16}=1$$

$$4(x+3)^2+(y-1)^2=16 \qquad \text{Multiplying by 16}$$

$$(y-1)^2=16-4(x+3)^2 \qquad \text{Subtracting } 4(x+3)^2 \text{ on both sides}$$

$$y-1=\pm\sqrt{16-4(x+3)^2} \qquad \text{Taking the square root on both sides}$$

$$y=1\pm\sqrt{16-4(x+3)^2} \qquad \text{Adding 1 on both sides}$$

Now we can use this equation to produce the graph in either of two ways. One is to enter $Y_1 = 1 + \sqrt{16-4(x+3)^2}$ and $Y_2 = 1 - \sqrt{16-4(x+3)^2}$, select a square window, and press $\boxed{\text{GRAPH}}$. Here we use [−9, 9, −6, 6]. The first equation produces the top half of the ellipse and the second equation produces the lower half.

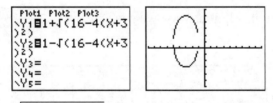

We can also enter $Y_1 = \sqrt{16-4(x+3)^2}$ and then enter $Y_2 = 1 + Y_1$ and $Y_3 = 1 - Y_1$.

Deselect Y_1, select a square window, and press $\boxed{\text{GRAPH}}$. We use Y_1 to eliminate the need to enter the radical expression more than once. Deselecting it prevents the graph of Y_1 from appearing on the screen with the graph of the ellipse. The top half of the graph is produced by Y_2 and the lower half by Y_3.

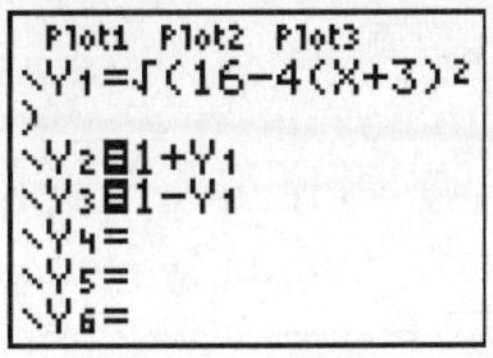

```
Plot1 Plot2 Plot3
\Y1=√(16-4(X+3)²
)
\Y2◻1+Y1
\Y3◻1-Y1
\Y4=
\Y5=
\Y6=
```

We could also begin by using the quadratic formula to solve the original equation for y.

$$4x^2 + y^2 + 24x - 2y + 21 = 0$$

$$y^2 - 2y + \left(4x^2 + 24x + 21\right) = 0$$

$$y = \frac{-(-2) \pm \sqrt{(-2)^2 - 4(1)\left(4x^2 + 24x + 21\right)}}{2(1)}$$

$$y = \frac{2 \pm \sqrt{4 - 16x^2 - 96x - 84}}{2}$$

$$y = \frac{2 \pm \sqrt{-16x^2 - 96x - 80}}{2}$$

Then enter $Y_1 = \dfrac{2 + \sqrt{-16x^2 - 96x - 80}}{2}$ and $Y_2 = \dfrac{2 - \sqrt{-16x^2 - 96x - 80}}{2}$ or enter

$Y_1 = \sqrt{-16x^2 - 96x - 80}$, $Y_2 = \dfrac{2 + Y_1}{2}$ and $Y_3 = \dfrac{2 - Y_1}{2}$. Then deselect Y_1, select a square

window and press GRAPH to display the graph shown on page 68.

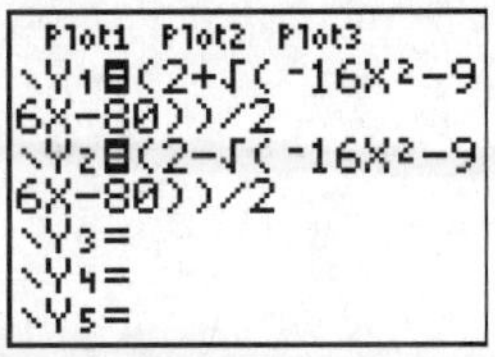

```
Plot1 Plot2 Plot3
\Y1◻(2+√(-16X²-9
6X-80))/2
\Y2◻(2-√(-16X²-9
6X-80))/2
\Y3=
\Y4=
\Y5=
```

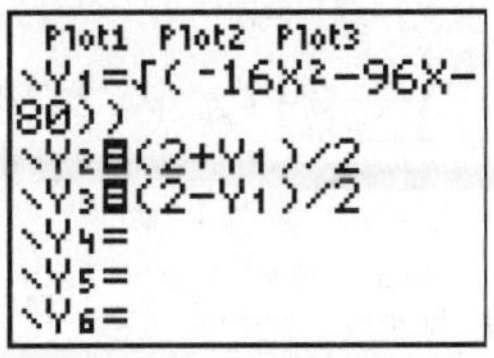

```
Plot1 Plot2 Plot3
\Y1=√(-16X²-96X-
80))
\Y2◻(2+Y1)/2
\Y3◻(2-Y1)/2
\Y4=
\Y5=
\Y6=
```

To use the Conics App to graph the ellipse, first press APPS, scroll down to Conics, press ENTER, and then press 2 to select the Ellipse menu. That menu displays the standard forms for the equation of an ellipse with center at (H, K). The first form is for an ellipse with a horizontal major axis and the second is for an ellipse with a vertical axis.

Our equation can be written as $\dfrac{\left[x - (-3)\right]^2}{2^2} + \dfrac{(y-1)^2}{4^2} = 1$, so this ellipse has a vertical

major axis. Thus, we select option 2, $\dfrac{(X-H)^2}{B^2} + \dfrac{(Y-K)^2}{A^2} = 1$. We have A = 4, B = 2,

H = −3, and K = 1.

Enter these values and then press GRAPH to see the graph in a window selected by the calculator. If you want the window to be automatically selected, be sure that WINDOW SETTINGS on the CONIC SETTINGS screen (obtained by pressing MODE) is set to AUTO. We can also select the window manually as described on page 65 of this manual.

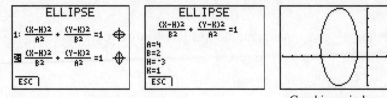

Graphing window
automatically selected

Quit the Conics App by pressing 2nd [QUIT].

GRAPHING HYPERBOLAS

As with equations of circles, parabolas, and ellipses, equations of hyperbolas must be solved for y before they can be entered on the equation-editor screen. They can also be graphed using the Conics App, if it is available.

Section 7.3, Example 2   Graph the hyperbola $9x^2 - 16y^2 = 144$.

First solve the equation for y.

$$9x^2 - 16y^2 = 144$$

$$-16y^2 = -9x^2 + 144 \qquad \text{Subtracting } 9x^2 \text{ on both sides}$$

$$y^2 = \frac{-9x^2 + 144}{-16} = \frac{9x^2 - 144}{16} \qquad \text{Dividing by } -16; \text{ factoring out } -1$$

$$y = \pm\sqrt{\frac{9x^2 - 144}{16}} \qquad \text{Taking the square root on both sides}$$

It is not necessary to simplify further.

Now enter $Y_1 = \sqrt{\dfrac{9x^2 - 144}{16}}$ and either $Y_2 = -\sqrt{\dfrac{9x^2 - 144}{16}}$ or $Y_2 = -Y_1$, select a square window, and press GRAPH. Here we use the window $[-9, 9, -6, 6]$. The top half of the graph is produced by Y_1 and the lower half by Y_2.

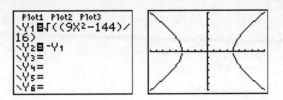

To use the Conics App to graph this hyperbola, first press $\boxed{\text{APPS}}$, scroll down to Conics, press $\boxed{\text{ENTER}}$, and then press $\boxed{3}$ to display the Hyperbola menu. In the text we see that our equation can be written in the form $\dfrac{x^2}{4^2} - \dfrac{y^2}{3^2} = 1$, so we select option 1. We have A = 4, B = 3, H = 0, and K = 0. Enter these values and then press $\boxed{\text{GRAPH}}$ to see the graph in a window selected by the calculator, assuming AUTO mode is selected. The window can be selected manually as described on page 65 of this manual.

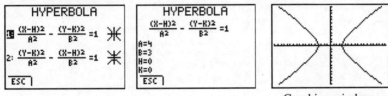

Graphing window
automatically selected

Quit the Conics App by pressing $\boxed{\text{2nd}}$ $\boxed{\text{QUIT}}$.

Section 7.3, Example 3 Graph the hyperbola $4y^2 - x^2 + 24y + 4x + 28 = 0$.

In the text we completed the square to get the standard form of the equation. Now solve the equation for y.

$$\frac{(y+3)^2}{1} - \frac{(x-2)^2}{4} = 1$$

$$(y+3)^2 = 1 + \frac{(x-2)^2}{4} \qquad \text{Adding } \frac{(x-2)^2}{4} \text{ on both sides}$$

$$y + 3 = \pm\sqrt{1 + \frac{(x-2)^2}{4}} \qquad \text{Taking the square root on both sides}$$

$$y = -3 \pm \sqrt{1 + \frac{(x-2)^2}{4}} \qquad \text{Subtracting 3 on both sides}$$

This equation can be used to produce the graph in either of two ways. One is to enter

$$Y_1 = -3 + \sqrt{1 + \frac{(x-2)^2}{4}} \text{ and } Y_2 = -3 - \sqrt{1 + \frac{(x-2)^2}{4}}, \text{ select a square window, and press}$$

GRAPH . Here we use [–12, 12, –9, 9]. The first equation produces the top half of the hyperbola and the second the lower half.

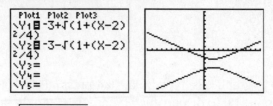

We can also enter $Y_1 = \sqrt{1 + \dfrac{(x-2)^2}{4}}$, $Y_2 = -3 + Y_1$, and $Y_3 = -3 - Y_1$. Deselect Y_1, select a square window, and press GRAPH . Again, Y_1 is used to eliminate the need to enter the radical expression more than once. Deselecting it prevents the graph of Y_1 from appearing on the screen with the graph of the hyperbola. The top half of the graph is produced by Y_2 and the lower half by Y_3.

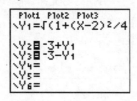

If the Conics App is available, it can also be used to graph this hyperbola. Access the Hyperbola menu in the Conics App as described earlier in this section on graphing

hyperbolas. We have $\dfrac{(y+3)^2}{1} - \dfrac{(x-2)^2}{4} = 1$ or $\dfrac{[y-(-3)]^2}{1^2} - \dfrac{(x-2)^2}{2^2} = 1$, so we select

option 2 and enter A = 1, B = 2, H = 2, and K = –3. Finally press GRAPH to see the graph in a window selected by the calculator, assuming AUTO mode is selected. We can also select a window manually as described on page 65 of this manual.

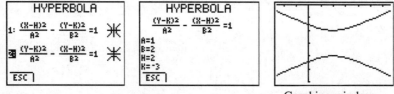

Graphing window
automatically selected

Quit the Conics App by pressing 2nd [QUIT].

Chapter 8 Sequences, Series, and Combinatorics

The computational capabilities of a graphing calculator can be used when working with sequences, series, and combinatorics.

FINDING THE TERMS OF A SEQUENCE

Section 8.1, Example 2 Use a graphing calculator to find the first 5 terms of the sequence whose general term is given by $a_n = n/(n+1)$.

Although we could use a table, we will use the Seq feature. The calculator can be set in either Func or Seq mode when this feature is used. Make sure that STAT WIZARDS in the MODE menu is ON. (See page 1 in this manual.) Press [2nd] [LIST] [▶] to display the LIST OPS menu. (LIST is the second operation associated with the [STAT] key.) Press [5] to open the "seq" screen. Then enter the general term of the sequence, replacing n with x if the calculator is in FUNC mode. Follow this with the variable and the numbers of the first and last terms desired. Press [X,T,Θ,n] [÷] [(] [X,T,Θ,n] [+] [1] [)] [,] [X,T,Θ,n] [,] [1] [,] [5] [,] [1] [ENTER].

The step option gives the ability to select values ranging from the starting value to the ending value in chosen increments. The default increment is 1.

If the calculator is in FUNC mode, pressing [X,T,Θ,n] will produce the variable x in the expression. If it is in SEQ mode the variable will be n.

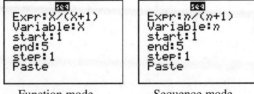

Function mode   Sequence mode

After [ENTER] has been pressed, the following screen appears.

Sequence mode, MathPrint mode
Use the right and left arrow keys to scroll between the screens.

Now press $\boxed{\text{MATH}}$ $\boxed{1}$ $\boxed{\text{ENTER}}$ to use the ▶Frac feature to express the terms as fractions.

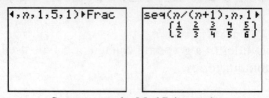

Sequence mode, MathPrint mode

If STAT WIZARDS is OFF, then "seq(" is pasted to the home screen when LIST OPS seq(is selected. Then the entries are entered by pressing $\boxed{\text{X,T,}\Theta\text{,}n}$ $\boxed{\div}$ $\boxed{(}$ $\boxed{\text{X,T,}\Theta\text{,}n}$ $\boxed{+}$ $\boxed{1}$ $\boxed{)}$ $\boxed{,}$ $\boxed{\text{X,T,}\Theta\text{,}n}$ $\boxed{,}$ $\boxed{1}$ $\boxed{,}$ $\boxed{5}$ $\boxed{,}$ $\boxed{1}$ $\boxed{)}$ $\boxed{\text{ENTER}}$. Note that the last entry, 1, is not necessary if you wish to use the default increment value. We include it here for completeness. We show the screen below in Classic mode to avoid scrolling to view the entire entry. Note that the variable is x, so the calculator is in FUNC mode.

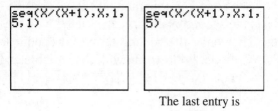

The last entry is
omitted here.

FINDING PARTIAL SUMS

We can use a graphing calculator to find partial sums of a sequence when a formula for the general term is known.

Section 8.1, Example 6 Use a graphing calculator to find S_1, S_2, S_3, and S_4 for the sequence whose general term is given by $a_n = n^2 - 3$.

We will use the cumSum feature. The calculator will write the partial sums as a list. First press $\boxed{\text{2nd}}$ $\boxed{\text{LIST}}$ $\boxed{\blacktriangleright}$ $\boxed{6}$ to paste "cumSum(" to the home screen. Then press $\boxed{\text{2nd}}$ $\boxed{\text{LIST}}$ $\boxed{\blacktriangleright}$ $\boxed{5}$ to paste "seq(" into the cumSum expression. If STAT WIZARDS is ON, then follow the procedure described on page 73, otherwise follow the procedure described above by pressing $\boxed{\text{X,T,}\Theta\text{,}n}$ $\boxed{x^2}$ $\boxed{-}$ $\boxed{3}$ $\boxed{,}$ $\boxed{\text{X,T,}\Theta\text{,}n}$ $\boxed{,}$ $\boxed{1}$ $\boxed{,}$ $\boxed{4}$ $\boxed{,}$ $\boxed{1}$ $\boxed{)}$ $\boxed{)}$ $\boxed{\text{ENTER}}$. As before, the last entry, 1, is not necessary if you wish to use the default increment value. We include it here for completeness. We show the result with the calculator set in SEQ mode. The variable will be expressed as x if the calculator is in FUNC mode.

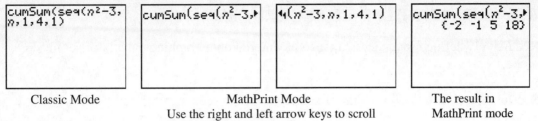

Classic Mode

MathPrint Mode
Use the right and left arrow keys to scroll
between the screens.

The result in
MathPrint mode

The TI-84 Plus can also compute individual partial sums. This is demonstrated in the next example.

Section 8.1, Example 7 (a) Evaluate $\displaystyle\sum_{k=1}^{5} k^3$.

The TI-84 has a summation function in the MATH MATH menu. Press MATH 0 to obtain the summation template. Press X,T,Θ,n 1 ▶ 5 ▶ X,T,Θ,n MATH 3 (or ^ 3) ▶ ENTER We can also use a shortcut to choose the summation template. Press ALPHA [F2] 2 to use the shortcut.

There is another way to compute the sum. Press 2nd [LIST] ▶ ▶ 5 (to select "sum" from the LIST MATH menu) 2nd [LIST] ▶ 5 (to select "seq" from the LIST OPS menu) X,T,Θ,n MATH 3 , X,T,Θ,n , 1 , 5)) ENTER. We show the result on a calculator set in Seq mode. If STAT WIZARDS is ON, then follow the procedure described on page 73.

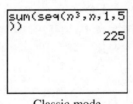

Classic mode

RECURSIVELY DEFINED SEQUENCES

Recursively defined sequences can also be entered on a graphing calculator set in SEQ mode.

Section 8.1, Example 9 Find the first 5 terms of the sequence defined by

$$a_1 = 5, \ a_{n+1} = 2a_n - 3, \text{ for } n \geq 1.$$

Press [Y=] and enter the recursive function beside "u(n) =" by pressing [2] [2nd] [u] [(]
[X,T,Θ,n] [−] [1] [)] [−] [3]. ("u" is the second operation associated with the [7] key.) Also set
u(nMin) = 5, the first term of the sequence. Press [2nd] [TBLSET] to display the TABLE
SETUP screen and set TblStart = 1, ΔTbl = 1, Indpnt: Auto, and Depend: Auto. Now
press [2nd] [TABLE] to display the table of values.

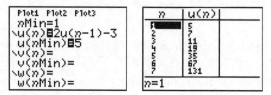

We see that $a_1 = 5$, $a_2 = 7$, $a_3 = 11$, $a_4 = 19$, and $a_5 = 35$.

EVALUATING FACTORIALS, PERMUTATIONS, AND COMBINATIONS

Operations from the MATH PRB (probability) menu can be used to evaluate factorials, permutations, and combinations.

Section 8.5, Example 3 (a) Compute $_4P_4$.

Press [4] [MATH] [▸] [▸] [▸] [2] [4] [ENTER]. These keystrokes enter 4, for 4 objects, display the
MATH PRB menu, select item 2, nPr, from that menu, enter 4, for 4 objects taken at a
time, and then cause the calculation to be performed. The result is 24, as shown in the
window on the next page.

Section 8.5, Example 6 Compute $_8P_4$.

Press [8] [MATH] [▸] [▸] [▸] [2] [4] [ENTER]. These keystrokes enter 8, for 8 objects, display the
MATH PRB menu, select item 2, nPr, from that menu, enter 4, for 4 objects taken at a
time, and then cause the calculation to be performed. The result is 1680, as shown on
next page.

```
4 nPr 4
            24
8 nPr 4
          1680
```

Section 8.5, Exercise 6 Evaluate 7!.

Press ⑦ MATH ▶ ▶ ▶ ④ ENTER. These keystrokes enter 7, display the MATH PRB menu, select item 4, !, from that menu, and then cause 7! to be evaluated. The result is 5040, as shown in the window below.

Section 8.5, Exercise 9 Evaluate $\dfrac{9!}{5!}$.

Press ⑨ MATH ▶ ▶ ▶ ④ ÷ ⑤ MATH ▶ ▶ ▶ ④ ENTER. The result is 3024.

```
7!
          5040
9!/5!
          3024
```

Section 8.6, Example 2 Evaluate $\begin{pmatrix} 7 \\ 5 \end{pmatrix}$.

Press ⑦ MATH ▶ ▶ ▶ ③ ⑤ ENTER. These keystrokes enter 7, for 7 objects, display the MATH PRB menu, select item 3, nCr, from that menu, enter 5, for 5 objects taken at a time, and then cause the calculation to be performed. The result is 21.

```
7 nCr 5
            21
```

Index

Absolute value, 2
Addition
 of complex numbers, 27
 of matrices, 50
Apps
 Conics, 63, 66, 69, 71, 72
 Inequalz, 58, 60, 61
 PolySmlt2, 49
 TVM Solver, 5
Ask mode for a table, 12
Auto mode for a table, 11

Base, change of, 37, 38

Catalog, 3, 17, 39
Change of base formula, 37, 38
Circles, graphing, 11, 65, 66
Classic mode, i–ii, 2
Combinations, 77
Complex numbers, 27
Composition of functions, 26
Conics App, 64, 67, 69, 71, 72
Contrast, 1
Converting to a fraction, 7
Coordinates of vertices, 61
Copying regression equation to
 equation-editor screen, 16, 17
Cramer's rule, 56
Cube root, 7
Cubic regression, 32
Curve fitting
 cubic regression, 32
 exponential regression, 40
 linear regression, 14
 logarithmic regression, 41
 logistic regression, 42
 quadratic regression, 31
 quartic regression, 32

Data, entering, 14
Deselecting an equation, 63
Determinants, 55
Diagnostic On/Off, 17

Division of complex numbers, 29
Dot mode, 57
DRAW menu, 11, 35, 57

e^x, 35
Editing entries, 6
Ellipses, graphing, 67–70
Entering data, 14
Equations
 graphing, 9
 solving, 19, 20
Evaluating functions, 12–13
Exponential functions, 35
Exponential regression, 39–40

Factorials, 77
Frac operation, ii, 7
Fraction, converting to, 8
Fraction shortcut, 5, 8
Fraction template, i, 5, 8
Function values, 12, 13
 maximum, 23
 minimum, 24
Functions
 composition, 26
 evaluating, 12–13
 exponential, 35
 graphing, 13
 greatest integer, 25
 inverse, 35
 logarithmic, 36
 maximum value, 23
 minimum value, 24
 zeros, 20–21

Graph styles
 dot, 57
 shade above, 59
 shade below, 57, 59
Graphing circles, 11, 65, 66
Graphing ellipses, 67–70
Graphing equations, 9
 points of intersection, 19

Graphing functions, 13
    inverse, 35
    logarithmic, 36
    rational, 33
Graphing hyperbolas, 70–72
Graphing inequalities, 57–60
Graphing an inverse function, 35
Graphing parabolas, 63–65
Greatest integer function, 25

Hyperbolas, graphing, 70–72

Inequalities, graphing, 23, 57–60
Inequalities, solutions of, 57–60
Inequalz App, 58, 60, 61
Intersect feature, 19
Intersection, points of, 19
Inverse function, graphing, 35
Inverse of a matrix, 54

Linear regression, 14
Ln x, 37
Log x, 36
Logarithmic functions, 33, 34
Logarithmic regression, 34
LogBASE, 37–38
Logistic regression, 42

MathPrint mode, i–ii, 1
Matrices, 45–56
    addition, 50
    determinant of, 45
    inverse of, 54
    multiplication, 53
    row-equivalent operations,
      47–49
    scalar multiplication, 52
    and solving systems of equations,
      45, 49, 55
    storing, 46
    subtraction, 51
Maximum function value, 23
Minimum function value, 24
Mode settings, 1
    dot, 57

Modeling
    cubic regression, 32
    exponential regression, 40
    linear regression, 14
    logarithmic regression, 41
    logistic regression, 42
    quadratic regression, 31
    quartic regression, 32
MTRX shortcut, ii, 46
Multiplication
    of complex numbers, 28
    of matrices, 53
    scalar, 52

Negative numbers, 2

On/Off, 1
Order of operations, 4

Parabolas, graphing, 63–65
Partial sums, 74
Permutations, 76
Piecewise functions, 24
Plotting points, 14–15
Points of intersection, 19
Points, plotting, 14–15
PoI-Trace, 61
PolySmlt2 App, 49

Quadratic regression, 31
Quartic regression, 30

Radical notation, 6
Rational exponents, 7
Rational functions, graphing, 33
Recursively defined sequences, 76
Regression
    cubic, 32
    exponential, 40
    linear, 14
    logarithmic, 41
    logistic, 42
    quadratic, 31
    quartic, 32
Regression equation, copying to
    equation-editor screen, 17–18

Roots of numbers, 7–8
Row-equivalent operations, 47–49

Scalar multiplication, 52
Scatterplot, 16
Scientific notation, 3
Sequences
        partial sums, 74,75
        recursively defined, 76
        terms, 73
Shading, 57
Shortcut menus, ii
        FRAC shortcut, 5
        FUNC shortcut, 3
        MTRX shortcut, 46
        YVARS shortcut, 16, 31
Solving equations, 19, 20
Square root, 6
Standard viewing window, 10
STAT lists, 14
STAT DIAGNOSTICS, 17
STAT PLOT, 15
        turning off, 9
STAT WIZARD, 16, 31,
        and exponential regression, 40
        and linear regression, 16
        and logarithmic regression, 41
        and logistic regression, 41
        and logarithmic regression, 41
        and partial sums, 74
        and terms of a sequence, 73
        turning off, 17
Storing values, 12
Subtraction
        of complex numbers, 27
        of matrices, 51
Summation, 75
Systems of equations
        and Cramer's rule, 45
        matrix solution, 45, 49
        and PolySmlt2 app, 49

Table feature
        ask mode, 11
        auto mode, 11
        and function values, 12–13
Terms of a sequence, 73
TEST menu, 24
Turning off plots, 9
TVM Solver, 5

Value feature, 13
Values of functions, 12, 13
VARS, 12, 16, 26
Vertices, coordinates, 60
Viewing window, 9
        standard, 10

Window, *see* Viewing window

x^{th} root, 7

Zero feature, 20
Zeros of functions, 20
ZDecimal (ZOOM 4), 33
ZoomStat (ZOOM 9), 16
Zstandard (ZOOM 6), 10